Self Assessment and Review of

STATISTICS

Dr. Sree Karuna Murthy Kolli

MD (CM), FAMS, FIPHA, FCGP, FAGE

Professor at Gayatri Vidya Parishad

Visakhapatnam

Peepee Publishers and Distributors (P) Ltd.®

Self Assessment and Review of Statistics

Published by
Pawaninder P. Vij and Anupam Vij

Peepee Publishers and Distributors (P) Ltd.
Head Office: 160, Shakti Vihar, Pitam Pura
Delhi-110 034 (India)

Correspondence Address:

7/31, First Floor, Ansari Road, Daryaganj
New Delhi-110002 (India)
Ph: 41512412, 23246245, 9811156083
e-mail: peepee160@yahoo.co.in
e-mail: peepee160@gmail.com
www.peepeepub.com

First Edition: **2019**

ISBN: 978-81-8445-253-2

Preface

On viewing the pattern of many DNB/NEET Exam patterns, I felt that there is no book available in the market for the best preparation to crack the exam by saving time of aspirants. ***Many of the exams had repeats all the time.***

This book is started with sole purpose of clearing various aspects of statistics and guide aspirants for their preparation. For controversial question every attempt has been made to trace the source and solve the questions.

I would like to concentrate all the contributors for their commitment and hard work.

Have faith in you, your parents and Almighty God.

Author
Karuna Murthy

Publisher
Pawaninder P. Vij and Anupam Vij

DISCLAIMER

Recent Pattern Questions

Book contains memory-based questions of NEET, DNB, NEET SS along with DNB SS and Institute based exams. ***Few questions are based on USMLE pattern which help students to gain extra knowledge.*** We don't claim that these questions are exact or similar to questions asked in recent examination. If any such similarity is found, it is purely coincidental and by chance.

No part of book should be copied or pasted any where. Publisher has every right to bring this to legal issue on concerned party for the same.

ABOUT THE PUBLISHER

I thank my publisher, **Mr. Pawaninder P. Vij and Mr. Anupam Vij**, Directors, **Peepee Publishers and Distributors (P) Ltd.**, who has always been with me and supported me. I would also like to thank his publishing team for their efforts in bringing this excellent book to this stage.

NOTICE:

Author and publisher has checked sources for their reliability to provide complete information and maintain standards needed. In spite of all these relentless efforts some errors might have crept in due to human error. Apart from that book is excellent review source of **STATISTICS.**

ABOUT THE EDITOR

Editor, **Kantamani Bala Teja,** is a renowned author of many superspeciality books like Mch Oncology and Mch Gastroenterology. He helped me to bring this book in an excellent manner by compiling necessary answers and helped me to solve controversial questions by consulting many statisticians.

Acknowledgements

There is a long list of people whom I wish to thank for their contributions. A simple thank you doesn't suffice to show my gratitude for their work.

I would like to thank the following persons for their contributions:

Dr. Kolli Bhaskari, MD, Community Medicine, GGH, Guntur
Dr. Ranga Rao, Vice Principal, GVP Hospital, Vizag
Dr. Archana Voleti, Microbiology
Dr. Rajani, Physiotherapy
Dr. Hanumantha Rao, Renowned Statistician

"He who makes us understand is not the teacher, but who moulds us to best is the one. These words played a crucial role in my life I expect same in yours too."

ONLY SHORT CUT FOR SUCCESS IN OUR EXAM IS HARD WORK.

ALL THE VERY BEST
Kantamani Bala Teja

WHY THIS BOOK??
• Contains Essential multiple choice questions of Statistics, Epidemiology and Screening
• All concepts are simplified keeping in view of understanding of students
• Excellent illustrations marked in tables and flowcharts
• All chapters are dealt in chapterwise with MCQs and summary at the end

FOR WHAT EXAMS THIS BOOK IS TO BE READ??
NEET/DNB PATTERN
• ***NAT board has changed its pattern***
• Frequently they are providing concept based MCQs rather than problem based questions
• So clear concepts of all topics are needed for NEET/DNB pattern exams
AIIMS PATTERN
• Every year one/two tricky question is made for AIIMS aspirants
• AIIMS design these MCQs only to eliminate competition
• TOPPERS are always at an extra edge by reading extra topics in a smart manner with long lasting memory
JIPMER PATTERN
• In spite of cut throat competition, JIPMER also creates new pattern of MCQs based on institute based protocols
• Research activity in this institute also creates new pattern of MCQs every year
USMLE PATTERN
• Very systematic and precise approach is needed for USMLE exam due to time constraint
• **Statistics and Epidemiology has an important role in cracking USMLE exams**
• Simplified concepts are mentioned in this book to answer them

A WORD ABOUT NEET SS AND DNB SS EXAMS

- Change in NEET SS and DNB SS pattern of exam had added 4 to 5 MCQs on this statistics part
- Book is also useful for DM/Mch/DNB pattern exams

HOW TO READ THIS BOOK?		
PG ASPIRANTS	**SUPERSPECIALITY ASPIRANTS**	**AIIMS/JIPMER ASPIRANTS**
Daily one chapter completes this book in 10-12 days	Read 1 chapter for every 10 days	Must read whole book 3 or 4 times before exam

Contents

PART 1: STATISTICS

PART 2: SCREENING

PART 3: EPIDEMIOLOGY

PART 4: MULTIPLE CHOICE QUESTIONS

PART 5: "THINK TANK"

PART 6: TABLES

PART 1: STATISTICS

CHAPTER 1

DATA

Data: Set of "Raw Facts / Values / Observations"

Example: Weight of 5 students in Class X = 42 kg, 45 kg, 48 kg, 51 kg, 53 kg

Information: "Processed and Meaningful Data"

Example: Average Weight of a student in Class X = 47.8 kg.

CLASSIFICATION OF DATA

- Data is classified into 2 types:

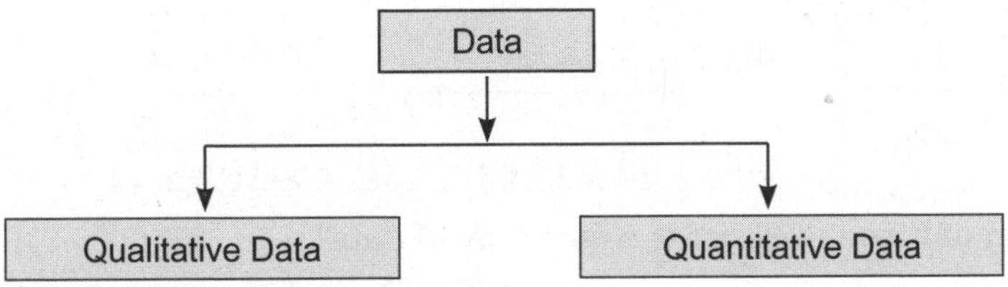

1. **Qualitative Data:**
   - Description of a character.
   - Categorizes the character.
   - May be Ordered or Unordered set of Values.
   - Equal Intervals between Values is ABSENT.
   - Answers to questions like–"What?", "Which?", "How do you Rate?"

Example:

- Sex–Male / Female
- Locality–Urban / Rural
- Blood Group–A / B / AB / O
- Family History of Cancer–Present / Absent
- Pain–Mild / Moderate / Severe
- Stage of Cancer–I / II / III / IV

2. **Quantitative Data:**
   - Numerical value of a Character.
   - Ordered set of Values.
   - Equal Intervals between Values are present.
   - Can either be COUNTED or MEASURED with a device.
   - Values are Whole Numbers or Decimal Numbers.
   - May have SI units of Measurements.
   - Answers to Q–"How Many?" / "How Much?"

> **Example:**
> - No. of children in a family, No. of Beds in a Ward
> - RBC Count (cells/ml), WBC Count (cells/ml)
> - Age (Yrs), Height (cms), Weight (Kg)
> - Blood Pressure (mm/Hg)
> - Blood Sugar level (mg/dl)
> - Serum LDL (mg/dl), Serum HDL (mg/dl)

MULTIPLE CHOICE QUESTIONS

1. **Measurement of blood pressure is which type of data?** **[NEET Pattern 2010]**
   (a) Nominal (b) Ordinal
   (c) Interval (d) Continuous

Ans: (d)
- Blood pressure is a quantitative, continuous and polyotomous variable.

2. **Blood groups will be an example for:** **[AIIMS 2004]**
   (a) Interval scale data (b) Ordinal scale data
   (c) Categorical data (d) None of the above

Ans: (c)
- Blood groups is an example of categorical data.

3. **Data is compiled based on presentation of patients with different blood groups for 1000 population. Which of the following is this type of data?** **[NEET SS Pattern]**
   (a) Nominal (b) Ordinal
   (c) Interval (d) None of the above

Ans: (a)
- There is no order–So it's not an ordinal data.
- There is no meaningful difference between variables–So it's not interval data.

SUMMARY

- **Nominal data** – Data divided into qualitative categories or groups, no order is seen. E.g. – Male/Female, Black/White.
- **Ordinal data** – Data has meaningful order but no information about size of interval.
- **Students raking in class** – 1st, 2nd, 3rd, Severity of illness – Mild, Moderate, Severe shows increasing severity. Like this meaningful order seen but no size difference.
- NOMINAL AND ORDINAL ARE QUALITATIVE DATA.
- **Interval** – Data has meaningful order and also meaningful interval. E.g., Temperature in deg C or Fahrenheit. No meaningful ratio.
- **Ratio** – Data has meaningful order and ratio also. E.g., Pulse rate 120/m is twice as fast as pulse rate of 60/m.

CHAPTER

2

VARIABLES

- Any attribute or any phenomenon or any event that can have different values.
- Most research questions are about the **relationship between two** or **more variables** in a specific group of individuals.

> **Examples:**
> - Relationship between **Smoking** and **Lung CA**
> - Relationship between **BMI** and **Diabetes**
> - Relationship between **Serum LDL level** and **BP**

CLASSIFICATION OF VARIABLE

- **Variables are classified into:**

DATA VARIABLE	
Type of Data	*Type of Variable*
Qualitative Data	Categorical variable
Quantitative Data	Discrete variable (Counts)
	Continuous variable (Measurements)

1. CATEGORICAL VARIABLES (QUALITATIVE):

- Description of a character.
- Answers to Questions: **"What?", "Which?", "How do you Rate?"**
- Can be either **Dichotomous or Polyotomous variables.**

> **Examples:**
> - Sex–Male/Female
> - Blood Group–A/B/AB/O
> - Pain–Mild/moderate/severe
> - Stage of Cancer–I/II/III/IV

2. DISCRETE VARIABLES (QUANTITATIVE):

- **Numerical value** of a Character.
- **Ordered** set of Values.
- **Equal Intervals** between Values are present.
- Can be **COUNTED**.
- Have only **FINITE NUMBER** of values.
- Values are always **Whole Numbers (No Decimal Points)**.
- They are all–**POLYOTOMOUS VARIABLES**.
- Answers to the Question–**"HOW MANY?"**

> **Examples:**
>
> – No. of children in family: 5, 1, 2, 6, 4
>
> – No. of episodes of Diarrhoea: 4, 7, 1, 8
>
> – No. of Beds in hospital wards: 20, 18, 26
>
> – Platelet count (cells/ml): 50000, 350000

3. CONTINUOUS VARIABLES (QUANTITATIVE):

- **Numerical value** of a Character.
- **Ordered** set of Values.
- **Equal Intervals** between Values are present.
- Can be **MEASURED** with a device.
- Values are **UNINTERRUPTED** (can have **DECIMAL POINTS**).
- Can have **INFINITE NUMBER** of possible values.
- They are all–**POLYOTOMOUS VARIABLES.**
- Answers to the Questions–**"HOW MUCH?"**

> **Examples:**
>
> – Weight of children (kg): 11, 9.8, 15.6
>
> – Height of Boys (meters): 1.87, 1.96, 1.45
>
> – BMI (kg/m^2): 25.6, 25.4, 29.3, 39.1
>
> – Hb (g/dl): 10.4, 12.0, 14.6, 6.2
>
> – Blood Sugar level (mg/dl): 133, 156, 220

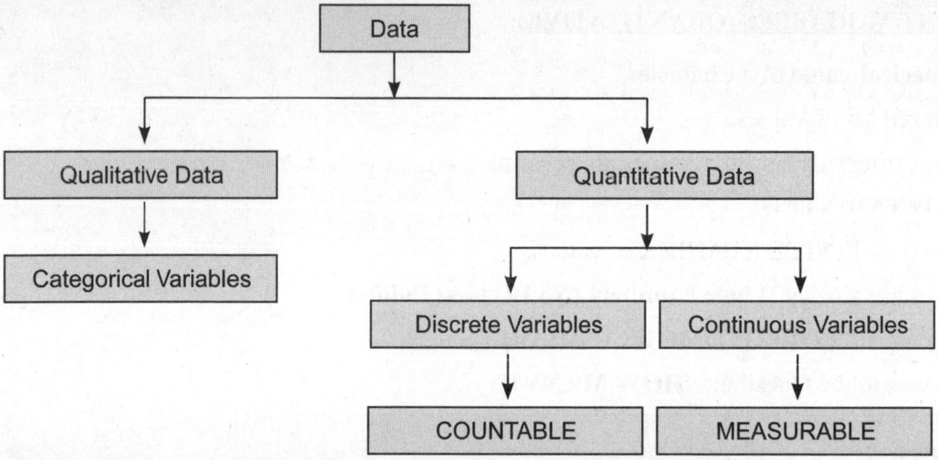

INDEPENDENT VARIABLES AND DEPENDENT VARIABLES:

Independent variables:	Dependent variables:
Precede dependent variables in time Often manipulated by the researcher Treatment/intervention that is used in a study	What is measured as an outcome in a study Values depend on the independent variable

MULTIPLE CHOICE QUESTIONS

1. All of the following are quantitative variables *except*: [NEET Pattern 2016]

(a) Serum cholesterol (b) Weight
(c) Gender (d) Celsius temperature scale

Ans: (c)

- **Gender**–Qualitative variable/Categorical variable.
- **Rest all are Quantitative.**

2. Platelet count is an example for:

(a) Categorical variable (b) Discrete variable
(c) Continuous variable (d) Dichotomous variable

Ans: (b)

- Platelet count is a Discrete variable as it is COUNT.

3. Serum Potassium level is an example for:

(a) Categorical variable (b) Discrete variable
(c) Continuous variable (d) Dichotomous variable

Ans: (d)

SUMMARY

TYPES OF VARIABLES AND THEIR CHARACTERISTICS:

Variable	Categorical	Discrete	Continuous
Data type	Qualitative	Quantitative	Quantitative
Character	Description	Numerical value	Numerical value
Data	Categories	Counts	Measurements
Values	Dichotomous or Polyotomous	Polyotomous	Polyotomous
Type of numbers	• No numbers, only categories • Can be converted into whole numbers for counting and analysis purpose	Finite numbers / Whole numbers (No decimal points)	Infinite numbers / decimal points
Answers to questions	"What?", "Which?" "How do you Rate?"	"How Many?"	"How Much?"
Examples	Blood Group–A/B/AB/O Pain–Mild/moderate/severe	No. of episodes of Diarrhoea: 4, 7, 10 Platelet count (cells/ml): 50000, 300000, 450000	BMI (kg/m²): 19.6, 23.4, 27.3, 32.1 Hb (g/dl): 11.4, 13.0, 14.6, 9.2

CHAPTER 3

SCALES

There are 4 scales of measurement of data:

- Nominal Scale
- Ordinal Scale
- Interval Scale
- Ratio Scale

DATA-VARIABLE-SCALE

Type of Data	Type of Variables	Type of Scale
Qualitative Data	Categorical	Nominal
		Ordinal
Quantitative Data	Discrete	Interval
	Continuous	Interval
		Ratio

1. **NOMINAL SCALE:**

   - **For CATEGORICAL VARIABLES (QUALITATIVE DATA).**
   - Set of categories identified only by **NAME.**
   - Classifies characteristics of people, objects or events into **CATEGORIES.**
   - Data has **NO specific Order.**
   - Data has **NO specific Interval.**

   Examples:
   - **Sex–Male/Female (Dichotomous variables)**
   - **Blood Group–A/B/AB/O (Polyotomous variables)**

2. **ORDINAL SCALE:**

   - **For CATEGORICAL VARIABLES (QUALITATIVE DATA).**
   - **Ordered** set of categories.
   - Data has a **specific Order or Ranking.**

- Shows the **direction** of difference between two individuals.
- **No specific interval** between two categories.

> **Examples:**
> - **Pain–Mild / Moderate / Severe**
> - **Stage of Cancer–I / II / III / IV**
> - **Likert Scale–Strongly Agree / Agree / Neutral / Disagree / Strongly Disagree**

3. **INTERVAL SCALE:**

- **For DISCRETE VARIABLES (QUANTITATIVE DATA).**
- **For CONTINUOUS VARIABLES (QUANTITATIVE DATA).**
- Has a **meaningful order.**
- Have equal intervals between the measurements.
- Data in Interval Scale **DO NOT HAVE AN ABSOLUTE ZERO.**

> **Examples:**
> - **IQ level**
> - **Year**
> - **Temperature in Fahrenheit (F) and Celsius (C°) scale**

4. **RATIO SCALE:**

- **For CONTINUOUS VARIABLES (QUANTITATIVE DATA).**
- Has a meaningful order.
- Have equal intervals between the measurements.
- Data in Ratio Scale HAVE AN ABSOLUTE ZERO.

> **Examples:**
> - **Height**
> - **Weight**
> - **Length**
> - **Temperature in Kelvin Scale**
> - **Blood Pressure**
> - **Blood Sugar level**
> - **Blood Cholesterol level**
> - **Hemoglobin level**

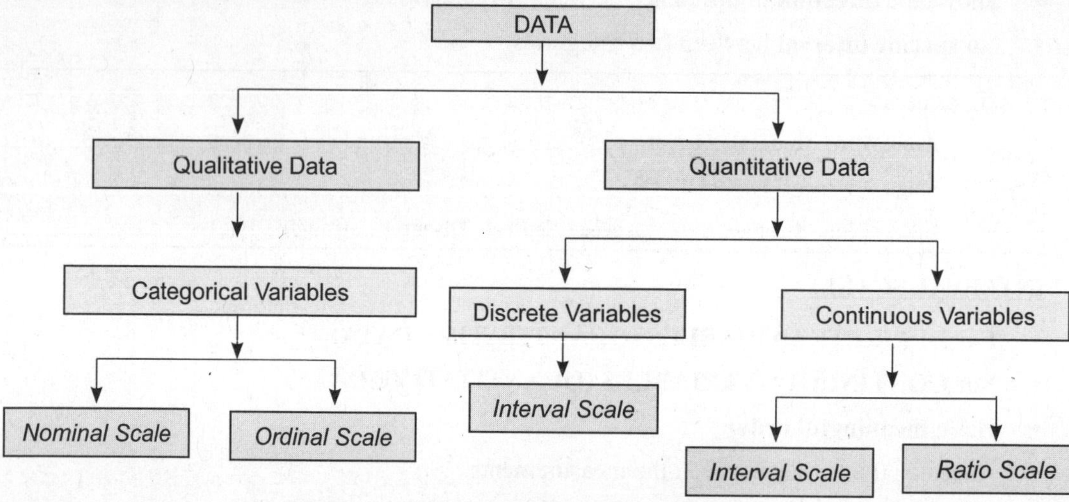

MULTIPLE CHOICE QUESTIONS

1. Which of the following variables is measured on the ordinal scale? [NEET 2009]

(a) Type of anemia
(b) Serum ferritin level
(c) Hemoglobin level
(d) Severity of anemia

Ans: (d)

- **Severity has an Order like mild, moderate and severe.**
- **Specified Order/Ranking.**
- So it is measured on Ordinal scale.

2. A physician, after examining a group of patients of a certain disease, classifies the condition of each one as 'Normal', 'Mild', 'Moderate' or 'Severe'. Which one of the following is the scale of measurement is adopted for classification of disease condition? [AIIMS 04]

(a) Normal
(b) Interval
(c) Ratio
(d) Ordinal

Ans: (d)

- Condition is classified as 'Normal', 'Mild', 'Moderate' or 'Severe'.
- So ordinal scale of measurement is to be adopted.
- **Specified Order/Ranking**–Ordinal scale is used.

3. Mean and standard deviation can be worked out only if data is on: [AIIMS 2003, 2005]

(a) Interval/Ratio scale
(b) Dichotomous scale
(c) Nominal scale
(d) Ordinal scale

Ans: (a)

- **For Quantitative data**–Interval/Ratio scale is used.

4. **An investigator into the life expectancy of IV drug abusers divides a sample of patients into HIV-positive and HIV-negative groups. What type of data does this division constitute? [AIIMS 2000]**

(a) Nominal (b) Ordinal
(c) Interval (d) Ratio

Ans: (a)

- Here there is no order and it cannot be measured directly.
- So, it can't be an **ordinal** data or **metric (Interval/ratio)** data respectively.

5. **Response which is graded by an observer on an agree/disagree continuum is based on: [NEET Pattern 2013]**

(a) Visual analog scale (b) Guttman scale
(c) Likert scale (d) Adjectival scale

Ans: (c)

> **Type of Ordinal Scale**
> Generally used to quantify attitudes and behavior
> Responses are graded on a continuum
> **Example:** Strongly agree – Agree – Neutral – Disagree – Strongly disagree

6. **In a study following interpretation are obtained: Satisfied, Very satisfied, Dissatisfied. Which type of scale is this? [AIIMS 2003]**

(a) Nominal (b) Ordinal
(c) Interval (d) Ratio

Ans: (b)

- **Specified Order/Ranking**–Ordinal scale is used.

7. **All of the following are example of nominal scale *except*: [NEET Pattern 2012]**

(a) Race (b) Sex
(c) Body weight (d) Socio-economic status

Ans: (c)

- Body weight is metric data.
- So it is not an example of NOMINAL.
- **NOMINAL = NAMES** ***"(Remember this students)"***.

8. **Likert scale is: [NEET Pattern 2016]**

(a) Ordinal scale (b) Nominal scale
(c) Variance scale (d) Categorical scale

Ans: (a)

- It is a Type of Ordinal scale.

9. Which type "Social Class" is, if it has four categories–I to V and Class I is the highest social class and Class V is the lowest? [NEET Pattern 2016]

(a) Dichotomous
(b) Nominal
(c) Ordinal
(d) Interval

Ans: (c)

- **Specified Order/Ranking**–Ordinal scale is used.

10. Among following Scale of data which is best measure of Quantitative variable? [NEET SS Pattern 2016]

(a) Nominal scale
(b) Ordinal scale
(c) Interval scale
(d) Ratio scale

Ans: (d)

- Ratio scale has all properties of interval variable and also has a zero point.
- So it is best as it has true zero point beyond which there are no values.

11. Scale used in measurement of presence or absence of risk factor is: [NEET SS Pattern 2016]

(a) Nominal scale
(b) Ordinal scale
(c) Interval scale
(d) Ratio scale

Ans: (a)

- It is qualitative data–So ordinal or nominal scales should be used.
- Since there is no specified order–Nominal scale is the best choice.
- Given data is a categorical data for which nominal data is used.

12. Scale used in measuring Serum creatinine is: [NEET SS Pattern 2016]

(a) Nominal scale
(b) Ordinal scale
(c) Interval scale
(d) Ratio scale

Ans: (d)

- Absolute zero point–It means there are no values beyond this point. No negative serum creatinine values.
- So it should be ratio scale.

13. Which of the following will be used for Qualitative assessment of perception based on summative scales?

(a) Guttmann scale
(b) Likert scale
(c) VAS scale
(d) Adjectival scale

Ans: (a)

GUTTMAN SCALE (CUMULATIVE SCALE):

- Contains a 'series of statements that **expresses increasing intensity**' of a characteristic AND respondent is asked to agree or disagree to with each statement.

- **Example:** Cadmium can cause Prostate cancer – Cadmium is an important cause of Prostate cancer – Cadmium is a very important cause of Prostate cancer and death – Cadmium is the most important cause of Prostate cancer and death in India).

SCALE	PARAMETER
Likert scale	Qualitative assessment of perception based on continuous/categorical scale
Guttmann scale	Qualitative assessment of perception based on discrete cumulative scales
Adjectival scale	Linguistic scale is a set of words of the same grammatical category

SUMMARY

Scale	Nominal	Ordinal	Interval	Ratio
Data	Qualitative	Qualitative	Quantitative	Quantitative
Variable	Categorical	Categorical	Discrete and Continuous	Continuous
Answers to Q	What? Which?	Ranking? Rating?	How Many? 'COUNTS'	How Much? 'MEASURES'
Values	Names	Rank/Rating	Whole Nos.	Decimal Nos.
Order	-	+	+	+
Equal Intervals	-	-	+	+
Absolute Zero	-	-	-	+
Example	Sex, Blood Group, Disease +/-	CA staging, SES class, Likert scale	No. of beds, WBC Count, Temp in F / C°	Ht, Wt, Hb%, RBS, BP, LDL, Temp in K

CHAPTER 4

REPRESENTATION OF DATA

Presentation of Data

Qualitative Data

Tables Diagrams:

i. Bar diagram
ii. Pie Chart
iii. Pictogram
iv. Spot Map
v. Shaded Map
vi. Cartogram

Quantitative Data

Tables Graphs:

i. Histogram
ii. Frequency Polygon
iii. Frequency curve
iv. OGIVE–Cumulative Frequency curve
v. Line Chart/Line Graph
vi. Scatter diagram
vii. Box and whisker plot
viii. Stem leaf plot

PART 1: PRESENTATION OF QUALITATIVE DATA

BAR DIAGRAM:

- **Use**: To compare the frequency distribution of **Categorical Variables (Qualitative Data).**
- Horizontal **X** axis (Abscissa) = **Categories of Variables.**
- Vertical **Y** Axis (Ordinate) = **Frequency (Number or %).**
- Length of bars = Frequency of the variables.
- Width of bars must be **Equal.**
- **Equal Space** or **Gap** is present between the bars.

TYPES OF BAR DIAGRAMS

A. Based on Direction of Bars:

i. Vertical Bar Diagram

ii. Horizontal Bar Diagram

B. Based on Type of Bars:

i. Simple Bar Diagram

ii. Multiple (or) Compound Bar Diagram

iii. Component and Proportional Bar Diagram

Vertical Bar Diagram

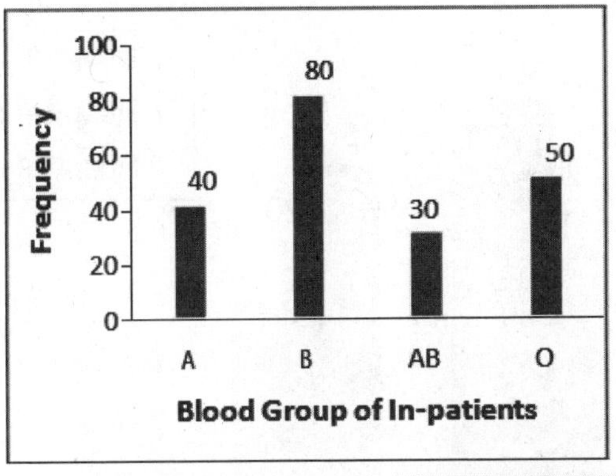

Horizontal Bar Diagram

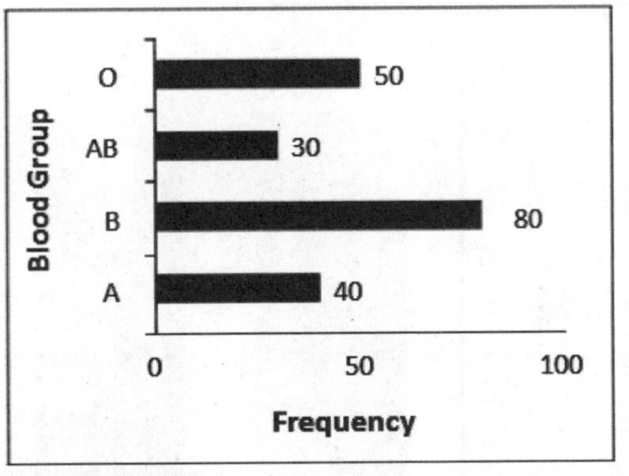

BASED ON TYPE OF BARS

I. SIMPLE BAR DIAGRAM:

- Depicts the frequency of Categories of a **Single Variable**.
- Consists of bars of a **single colour** or shade.
- **Equal Space between the bars is present.**

Example:

Distribution of In-Patients of a Hospital 'X' in a month based on their Blood Group

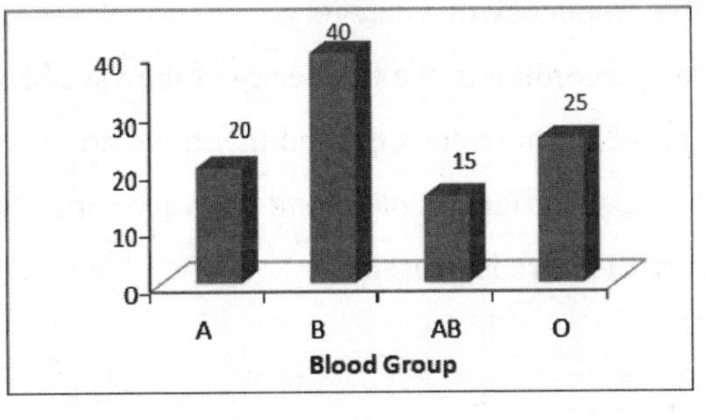

II. MULTIPLE / COMPOUND BAR DIAGRAM:

- **Use:** To compare frequency of ≥ 2 sub-groups within the Categories of a variable.
- Bars are placed adjacent to each other–**Adjacent Bar Diagram.**
- Bars are of Multiple colours or shades.
- Horizontal X axis = ≥ 2 groups within Categories.
- Vertical Y Axis = Frequency (Number or %).

Example:

Comparison of distribution of In-Patients of 3 Hospitals based on their Blood Group

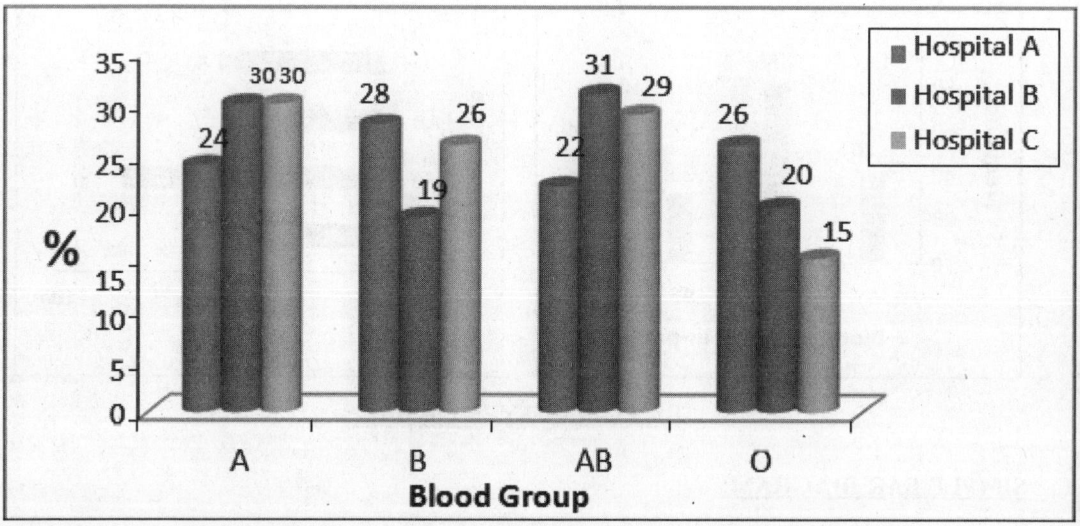

III. COMPONENT BAR DIAGRAM:

- Simple bar diagram externally and Multiple bar diagram internally.
- X axis: Categories of variables; Y axis: Frequency (Number).
- Length of bars = Total Frequency of a category.
- Height of the bars vary according to the frequency of the variables.
- Each bar is divided into 2 or more parts depending on the no. of sub-groups in a category.
- Each part within a bar has a different colour and so every bar is **Multicoloured**.
- **Equal Space between the bars is present.**

Example:

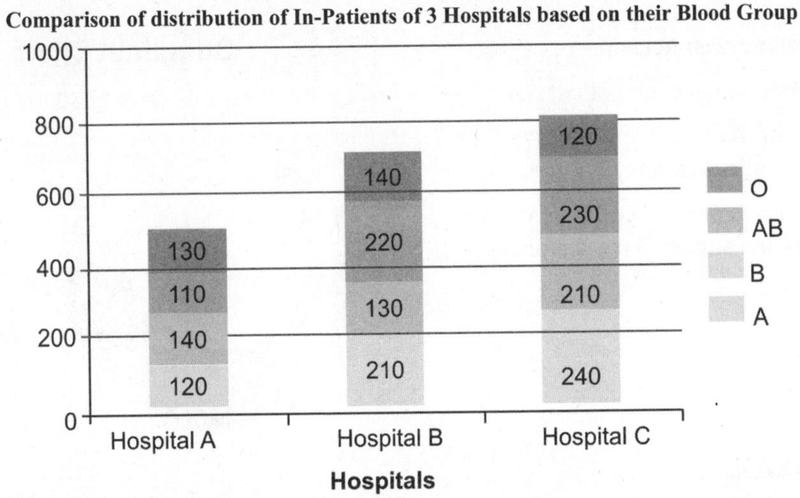

IV. PIE CHART / SECTOR DIAGRAM:

- Area of the sectors of the circle and their angulations are proportional to the frequency of categories of the variable.
- Each sector is coloured differently.
- All categories must be **Mutually Exclusive** (total must be 100% (= 360°)).
- **Cannot** be used to **compare** frequency distribution of 2 different variables in one pie chart.

Example:

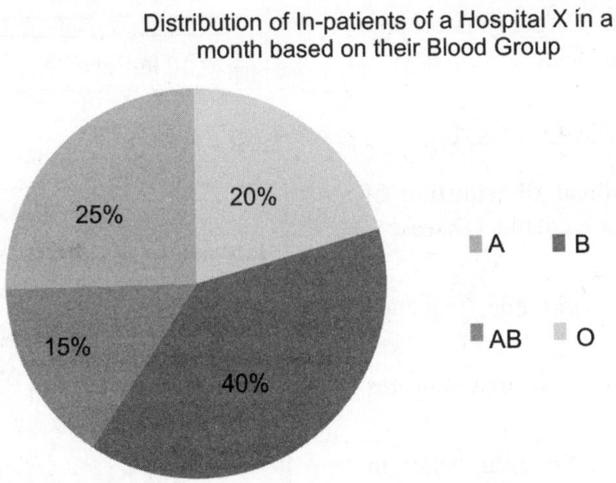

DOUGHNUT CHART

- It is essentially a pie chart.
- But an area of the center is cut out.
- **Length of Arcs** are proportional to frequency of categories of a variable.
- Length of arcs must be compared rather than area and angles of the sectors as in pie chart.

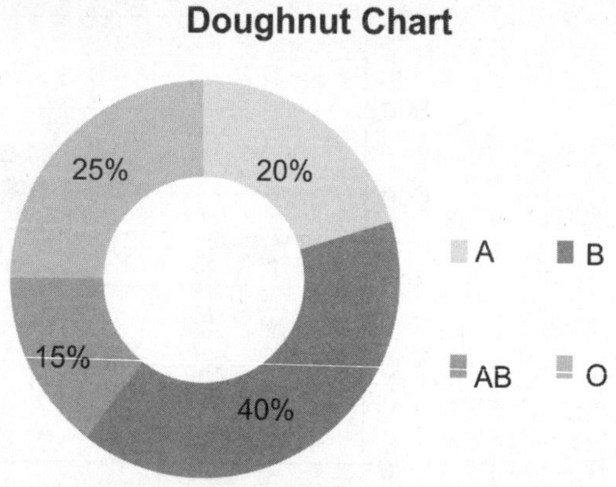

V. PICTOGRAM:

- **Images / Symbols** depict frequency of categories.
- LAY MAN CAN EASILY UNDERSTAND.
- X axis: Frequency; Y axis: Category (usually).
- Value assigned for an Image/Symbol is FIXED.
- No. of Images in Category is proportional to its frequency.

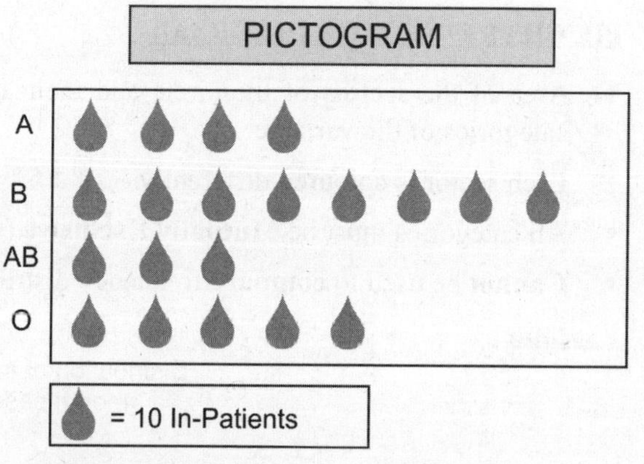

VI. SPOT MAP / MAP DIAGRAM:

- Shows **geographical distribution** of frequencies of a variable (Disease / Death).
- Each spot (dot) marks one frequency or its multiples.
- Clustering of spots in an area indicates common source.
- First developed by **Sir John Snow** in 1858.
- Investigated a **Cholera epidemic** in London.

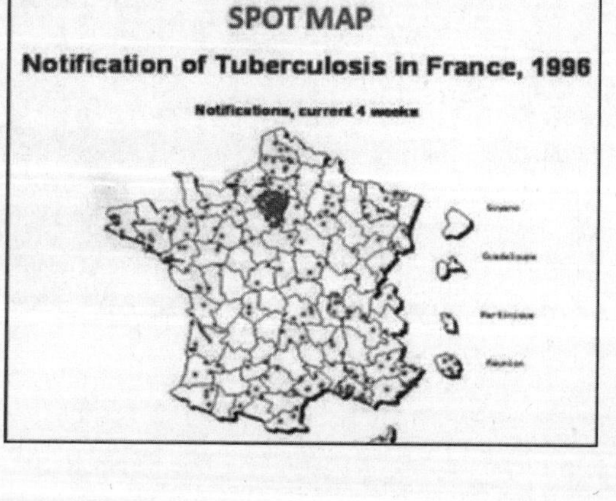

VII. SHADED MAP–CHOROCHROMATIC MAP:

- Differences in shading, colouring, or symbols indicates the character of a population or event or disease **(Presence/Absence (or) Low / Medium / High)** in different parts of the World / Country / State during a particular time (month or year).
- Legend explains the scale of the variable.

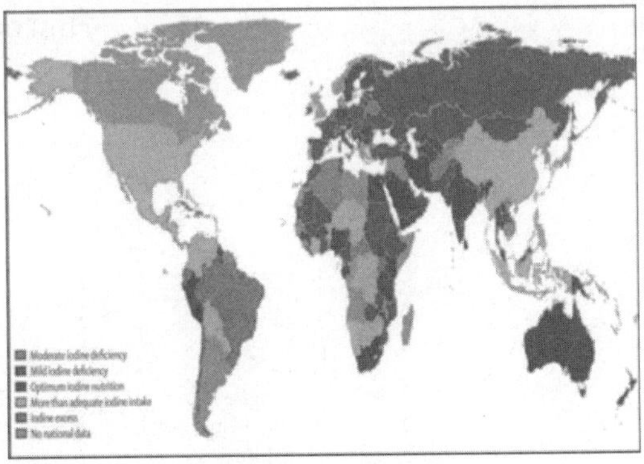

PART 2: PRESENTATION OF QUANTITATIVE DATA

I. FREQUENCY DISTRIBUTION TABLE:

Frequency Distribution Table

Frequency distribution of test marks of class 'A' students

Test Marks (Class Interval)	No. of Students Frequency (f)	Relative frequency (f/N)	%
70-74	8	0.16	16
75-79	8	0.16	16
80-84	5	0.10	10
85-89	11	0.22	22
90-94	9	0.18	18
95-99	5	0.10	10
100-104	4	0.08	8
Total	**N = 50**	**1.00**	**100**

II. HISTOGRAM:

- X axis : Class Intervals of the variable.
- Y axis : Frequency (Numbers).
- Consists of a series of **vertical bars–Uniform** in **colour.**
- Height of the bars = Frequency of a Class Interval.
- **NO GAPS** or **SPACE** between the Bars.
- **Can't** be used to **compare** frequency distribution of **2 different variables** in one histogram.
- Can depict whether the data is **Unimodal / Bi-modal / Multimodal.**

- Can depict whether the data is **Normally distributed** or **Skewed (+ve / –ve).**

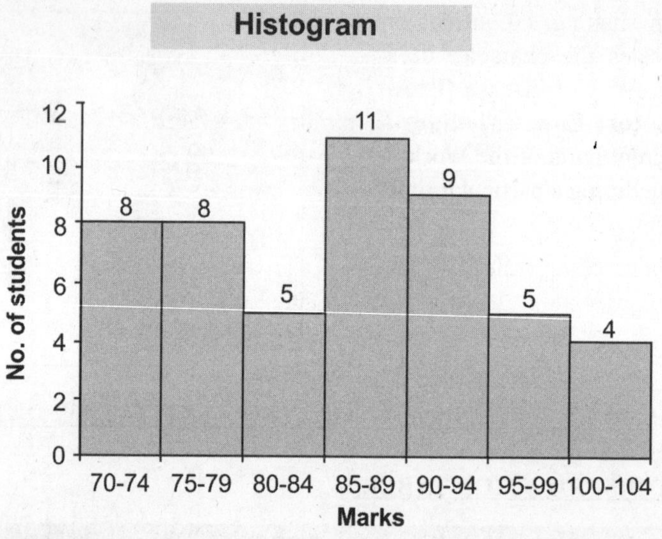

III. FREQUENCY POLYGON:

- **AREA DIAGRAM** of frequency distribution.
- Drawn by joining the **Midpoints** of all the class intervals of a HISTOGRAM by a line and the bars are removed.
- **X-axis:** Class-Midpoints.
- **Y-axis:** Frequency/Relative frequency.
- A Frequency Polygon **TOUCHES the X-axis** on both ends.
- Comparison of ≥ 2 groups of a continuous variable-possible.

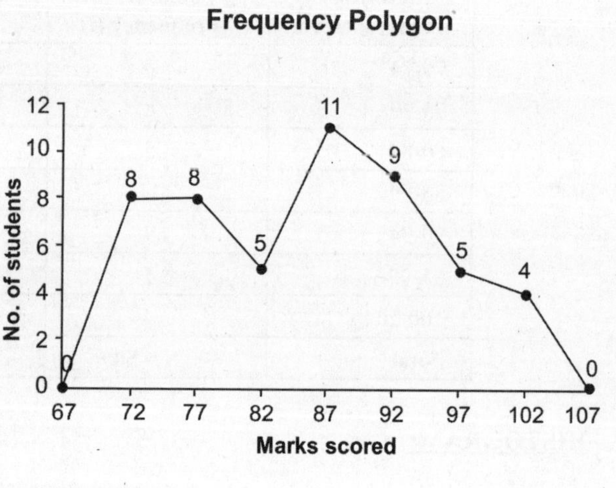

IV. FREQUENCY CURVE:

- When the No. of observations is large and width of **class intervals is reduced** – a Frequency Polygon loses its angulations to become a curve.
- A Frequency curve **TOUCHES the X-axis** on both ends.

- Depicts whether the data is Unimodal / Bi-modal / Multimodal.
- Depicts whether the data is normally distributed or Skewed (Positively / Negatively).

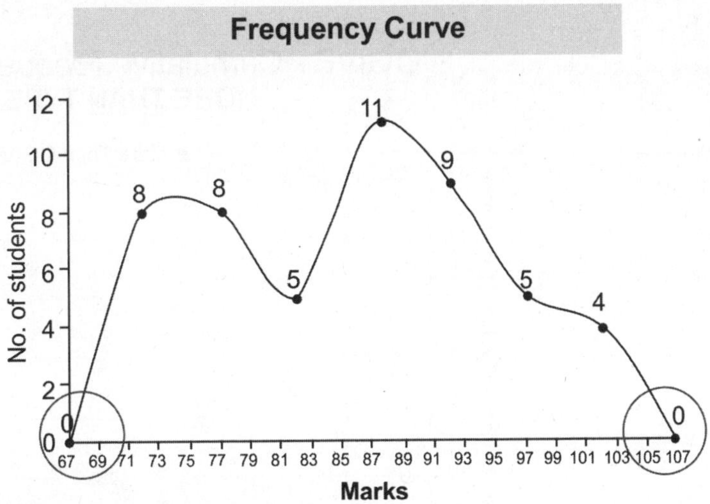

V. OGIVE (CUMULATIVE FREQUENCY CURVE):

- Graphical method for location of **Percentiles / Quartiles.**
- It is a graph of **Cumulative Frequency Distribution**.
- 2 types:

i. Less than OGIVE.

ii. More than OGIVE.

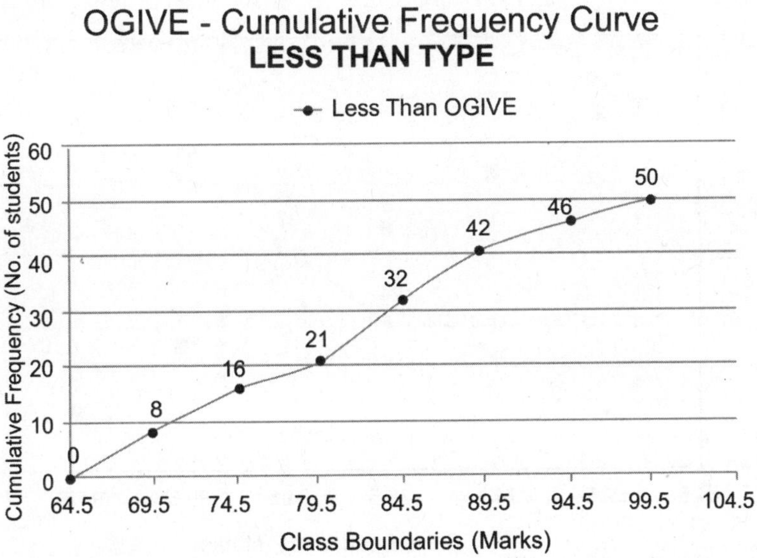

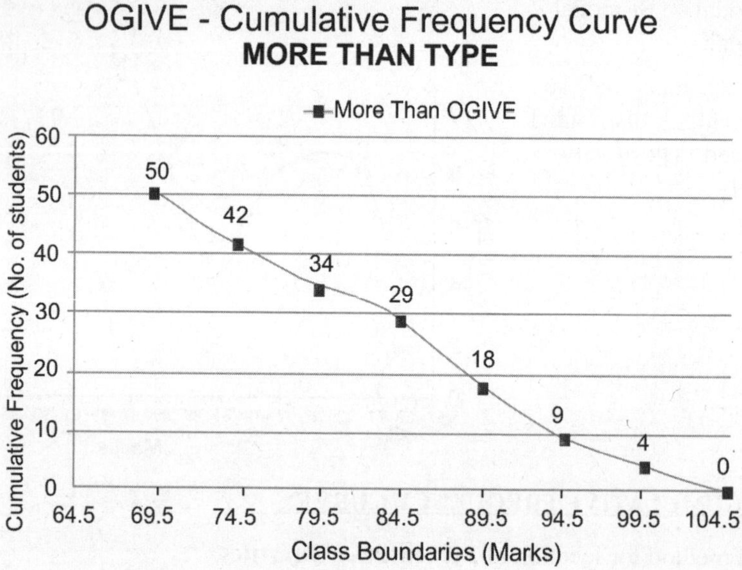
OGIVE - Cumulative Frequency Curve
MORE THAN TYPE
More Than OGIVE
Cumulative Frequency (No. of students)
0
10
20
30
40
50
60
50
42
34
29
18
9
4
0
64.5
69.5
74.5
79.5
84.5
89.5
94.5
99.5
104.5
Class Boundaries (Marks)

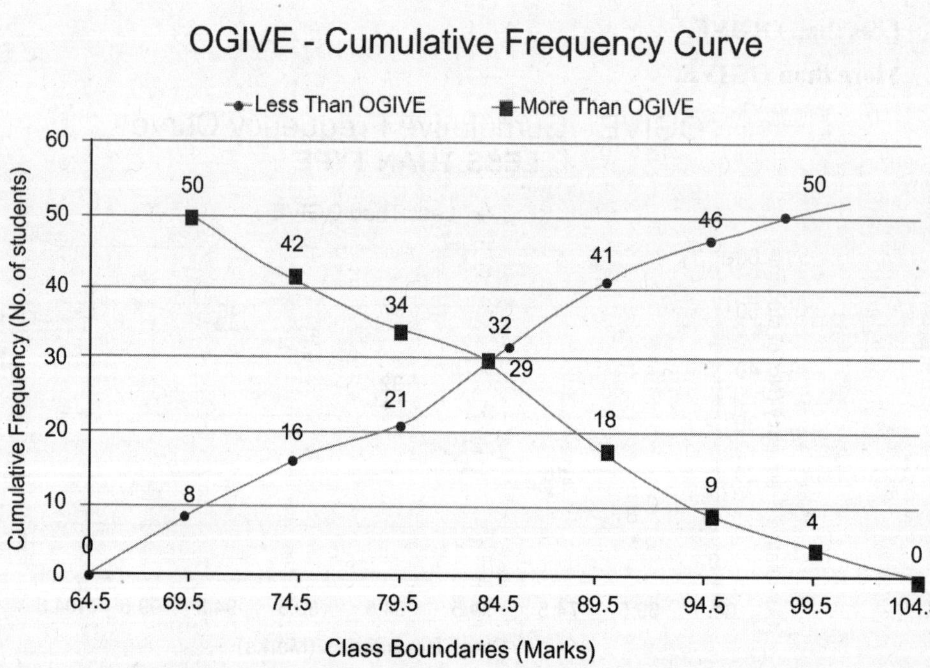
OGIVE - Cumulative Frequency Curve
Less Than OGIVE
More Than OGIVE
Cumulative Frequency (No. of students)
0
10
20
30
40
50
60
50
42
34
32
29
21
16
8
0
41
46
50
18
9
4
0
64.5
69.5
74.5
79.5
84.5
89.5
94.5
99.5
104.5
Class Boundaries (Marks)

VI. LINE DIAGRAM/LINE CHART/LINE GRAPH:

- It is a type of FREQUENCY POLYGON.
- **Use:** To show the **TREND OF EVENTS** over a period of time **(TIME TREND).**
- It presents variations of frequency by line.
- **X-axis:** TIME (Years/Months/Weeks/Days/Hrs/Mins).
- **Y-axis:** Frequency (Number).
- **Line DOES NOT touch the X-axis.**
- Comparison of time trend of ≥ 2 groups in a variable can be done.

Line Chart / Line Graph / Line Diagram

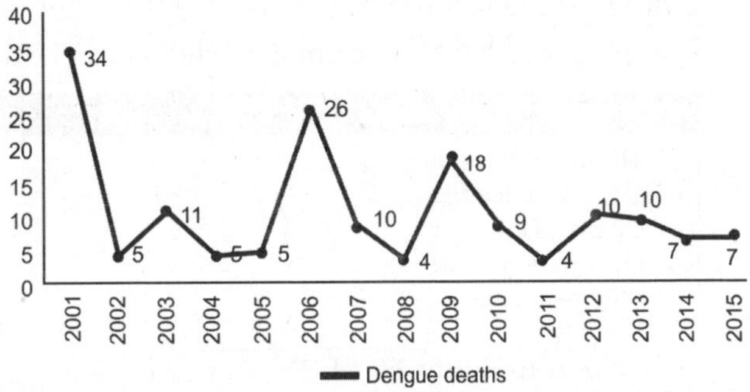

VII. SCATTER DIAGRAM / DOT DIAGRAM:

- **Use:** To depict the **Correlation** (relationship) between **2 continuous variables (Quantitative Data).**
- **X-axis:** Independent Variable or Exposure Variable.
- **Y-axis:** Dependent Variable or Outcome Variable.
- **Trend line** drawn along the distribution of dots / scatter, shows **Direction** of relationship.
- Also called as **"Correlation diagram".**

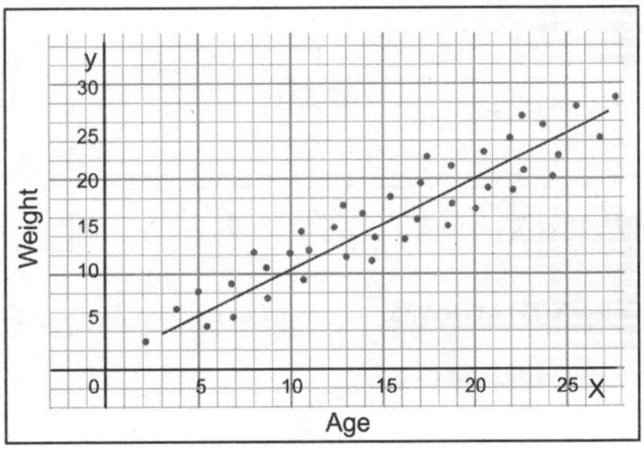

VIII. STEM LEAF PLOT:

- **Use:** To display **Quantitative Data.**
- Each value is divided into 2 parts–a stem and a leaf.
- The first part contains the **first digit = stem.**
- The second part contains the **second digit = leaf.**
- Shows trend of data with respect to variable.

STEM LEAF PLOT

STEM	LEAF
7	1 1 2 5 7 7
8	3 5 6 7 8 8
9	0 1 1 6 6 8 8
10	0

IX. BOX AND WHISKER PLOT:

- **Use:** To present **Quantitative Data.**
- Measured in Quartiles.
- Helps in General assessment of trend of data with respect to higher or lower values.
- Comparison of ≥ 2 **Quantitative Variables** can be done.

Five–Number Summary
– The maximum value
– Q3–Upper quartile
– Q2 (Median)
– Q1–Lower quartile
– Minimum value

BOX AND WHISKER PLOT

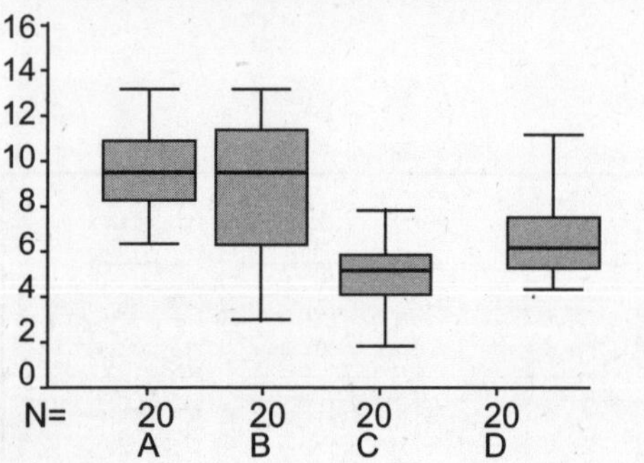

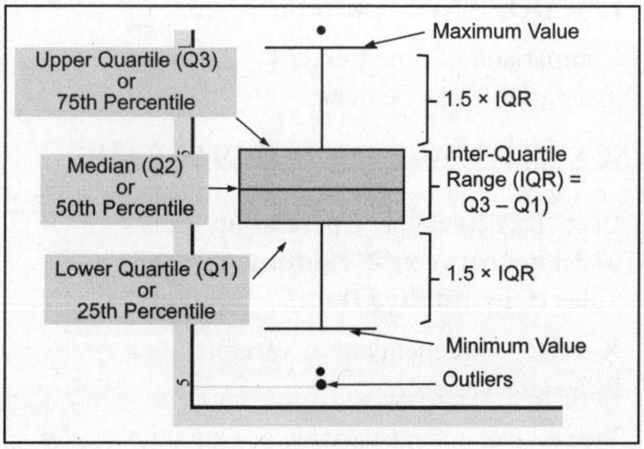

X. VENN DIAGRAM:

Represents logistic overlap

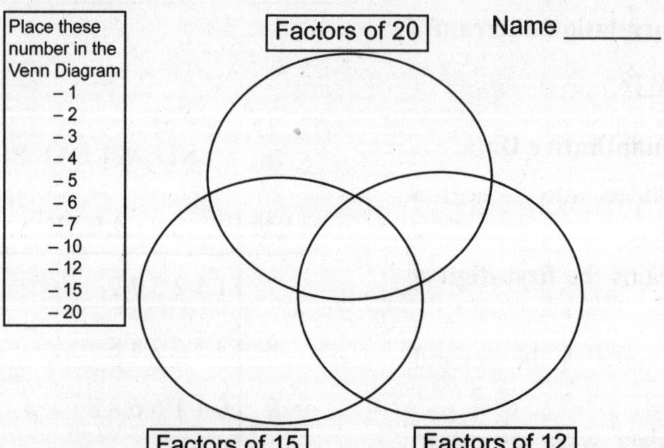

MULTIPLE CHOICE QUESTIONS

1. **Histogram is used to describe:** **[AIIMS 2005]**
   (a) Data collected on ordinal scale
   (b) Qualitative data of a group of patients
   (c) Data collected on nominal scale
   (d) Quantitative data of a group of patients

Ans: (d)

- Histogram is useful for graphical presentation for 'continuous quantitative data'.

2. **Histogram is used to present which kind of the data:** **[AIIMS May 2006]**
   (a) Nominal
   (b) Continuous
   (c) Discrete
   (d) Any of the above

Ans: (b)

- Histogram is useful for graphical presentation for 'continuous quantitative data'.

3. **Graph to correlate two quantitative data is:** **[NEET Pattern 2017]**
   (a) Histogram
   (b) Scatter diagram
   (c) Line diagram
   (d) Frequency curve

Ans: (b)

- Correlation or relation between 2 variables is represented by Scatter diagram.

4. **Best way to plot change of incidence of disease over time:** **[NEET Pattern 2012]**
   (a) Histogram
   (b) Line chart
   (c) Scatter diagram
   (d) Ogive

Ans: (b)

- Line chart represents trend of events and plots change of incidence over time.

5, **Graph showing relation between 2 variables is:** **[NEET Pattern 2011]**
   (a) Scatter diagram
   (b) Frequency polygon
   (c) Picture chart
   (d) Histogram

Ans: (a)

- Scatter diagram gives relation between 2 variables.
- Correlation diagram is synonym for it.

6. **Histogram is used as method of group presentation for:** **[NEET Pattern 2003]**
   (a) Qualitative data
   (b) Quantitative continuous data
   (c) Quantitative data–discrete type
   (d) Norminal data

Ans: (b)

- Histogram is useful for graphical presentation for 'continuous quantitative data'.

7. All of the following methods can show relationship between two variables *except*: **[NEET Pattern 2015]**

(a) Histogram
(b) Line diagram
(c) Bar chart
(d) Scatter plot

Ans: (c)

- Bar chart is only for visual comparison of different frequencies in discrete data.

8. A Scatter diagram is drawn to study: **[AIIMS 2007]**

(a) Trend of a variable over a period of time
(b) Frequency of occurrence of events
(c) Mean and median values of the given data
(d) Relationship between two given variables

Ans: (d)

- Scatter diagram gives relation between 2 variables.
- Correlation diagram is synonym for it.

9. Frequency is represented as an area in continuous pattern in: **[NEET Pattern 2014]**

(a) Bar diagram
(b) Histogram
(c) Pie diagram
(d) Pictogram

Ans: (b)

- Discrete continuous data is represented best by Histogram.

10. Best method to show trend of events with passage of time is: **[NEET Pattern 2014]**

(a) Line diagram
(b) Bar diagram
(c) Histogram
(d) Pie chart

Ans: (a)

- Line diagram represents trend of events with passage of time.

11. Trends can be best represented by: **[NEET Pattern 2013]**

(a) Scatter diagram
(b) Bar diagram
(c) Line diagram
(d) Pie chart

Ans: (c)

- Line diagram represents trend of events with passage of time.

12. Which of following represent frequency of continuous variables: **[NEET Pattern 2015]**

(a) Histogram
(b) Line diagram
(c) Simple bar chart
(d) Component bar chart

Ans: (a)

- Frequency of Continuous variables are represented best by Histogram.

13. Which of the following is used to denote a continuous variable? **[AIIMS 2010]**

(a) Simple bar
(b) Histogram
(c) Pie diagram
(d) Multiple bar

Ans: (b)

- Continuous variables are represented best by Histogram.

14. Best method to plot the Incidence of a disease over time is: **[NEET SS Pattern]**

(a) Histogram
(b) Line Chart
(c) Scatter Diagram
(d) Ogive

Ans: (b)

- Trend of events are described by Line chart.
- So Incidence is best plotted by Line chart.

15. Graph that is used to correlate two Quantitative data is: **[NEET SS Pattern]**

(a) Histogram
(b) Line Diagram
(c) Scatter Diagram
(d) Frequency curve

Ans: (c)

- Scatter diagram is called as Correlation diagram as it correlates 2 Quantitative data.

16. All of the following graphical methods can compare between two variables *except*: **[NEET SS Pattern]**

(a) Bar chart
(b) Line Diagram
(c) Histogram
(d) Scatter Diagram

Ans: (c)

- **Most versatile of all statistical diagrams is BAR DIAGRAM**
- Bar diagram–Visual comparison of magnitude of different frequencies in discrete data
- Scatter diagram is also called as Correlation diagram as it correlates 2 Quantitative data
- Line chart–Is a frequency polygon presenting variations by line

17. Numerical data presented in descriptive form is: **[NEET SS Pattern]**

(a) Graphical presentation
(b) Tabular presentation
(c) Textual presentation
(d) Diagrammatic presentation

Ans: (c)

- Textual presentation is for descriptive form.

18. Zig zag graph shows: **[NEET SS Pattern]**

(a) Components

(b) Fluctuations

(c) Median

(d) Correlation

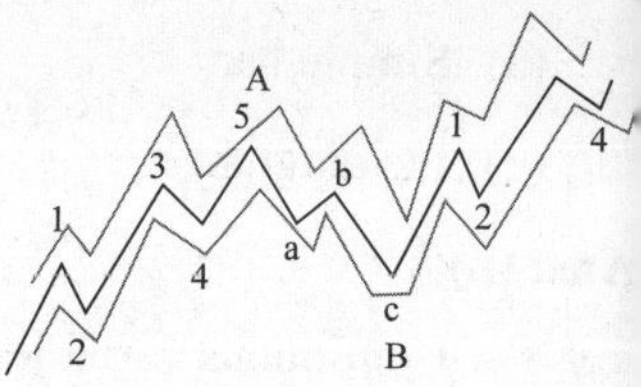

Ans: (b)

19. Which of the following is true about bar diagram: **[NEET SS Pattern]**

(a) Used for quantitative variables

(b) Distantly placed column chart

(c) Joint column chart

(d) Gives estimate for data variation

Ans: (c)

- **Bar diagram is for qualitative data**
- **Histogram is a column chart**
- **Bar diagram is a Joint column chart**

20. Pie chart is of following components *except*:

(a) Percentages

(b) Sectors

(c) Angles

(d) Bars

Ans: (d)

- Bars are best represented in Bar diagram.
- Rest all can be represented in Pie chart.

CHAPTER

5

STATISTICS AND ITS COMPONENTS

CLASSIFICATION:

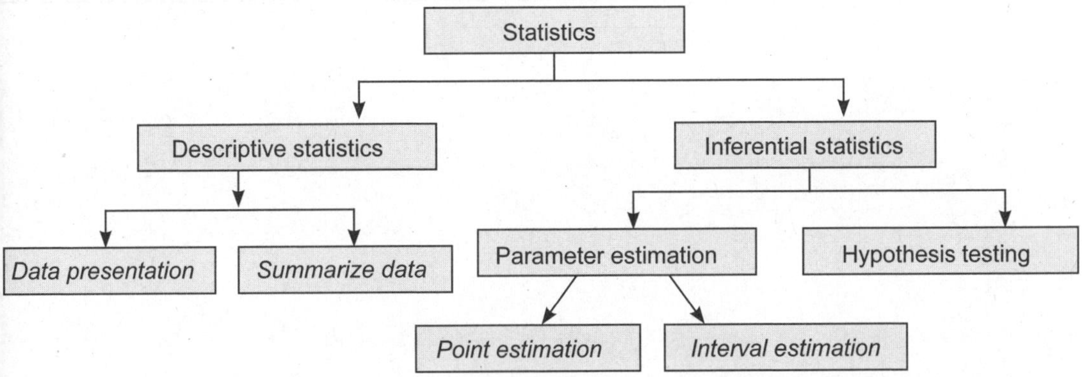

Descriptive Statistics	Inferential Statistics	Statistical Inference
Summarizes or Describes data set of sample. Evaluates the data set for patterns. Reduces information to convenient form.	Use sample data to make estimates or predictions about the larger population.	Use the data from the sample to inform us about the population (E.g., from sample mean X we estimate the population mean μ)

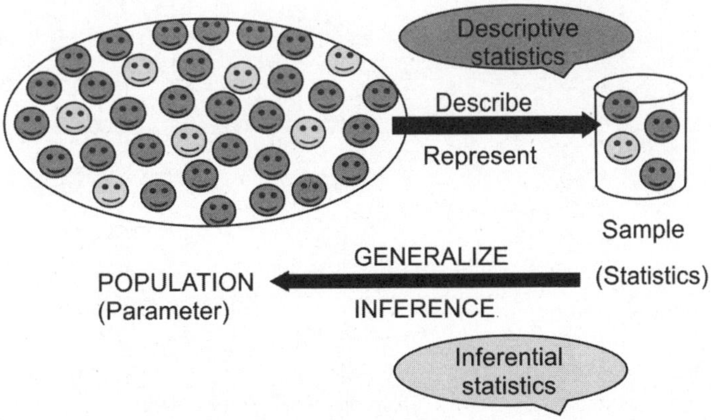

INFERENTIAL STATISTICS

TYPES:

Parametric Estimation	Hypothesis Testing
Point Estimation: Estimation of Mean, Proportion **Interval Estimation:** Confidence limits and Interval	Test of Significance

CHAPTER 6
MEASURES OF CENTRAL TENDENCY

These measures describe the location of the middle point of the data.

MEASURES OF CENTRAL TENDENCY		
Mean	Median	Mode

MEAN:

- Average of all the Values in a Data set.
- It is the sum of all the values divided by the total number of values in a data set.

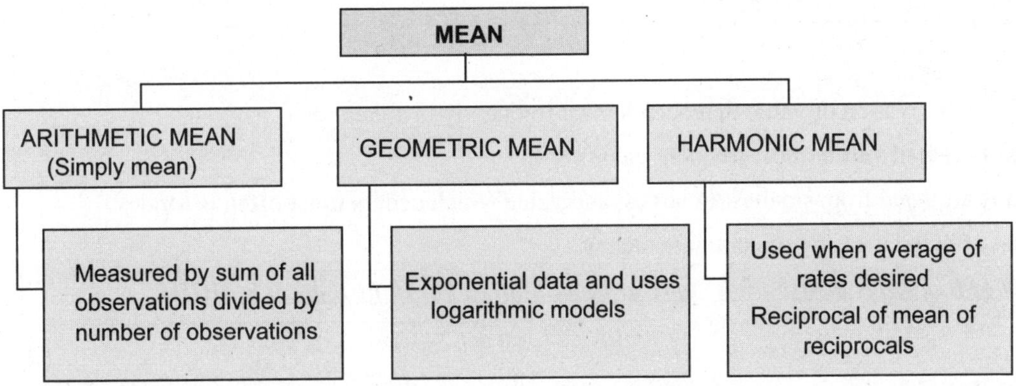

ADVANTAGES:

- All values are considered for calculation
- Mean is used for both continuous and discrete numeric data
- Mean is valid only for interval or ratio data

LIMITATIONS:

- Mean can't be calculated for categorical data
- Mean is being affected by an outlier
- **Most Affected** by extreme values in an asymmetric or skewed distribution

Ungrouped	Grouped
$x = \Sigma\frac{x}{n}$	$x = \Sigma\,\frac{(x, f)}{n}$

MEDIAN:

- **Middle most Value in a data set which is arranged in ascending or descending order.**
- It divides the arranged series of data in two equal parts.

> - If the data set has an odd number of entries, the median is the middle most value **((n + 1)/2)th value.**
> - If the data set has an even number of entries, the median is the average of the two middle values =**(n/2 and (n + 1)/2)th value.**

ADVANTAGES:

- Median can be determined by ordinal data as well as interval and ratio data.
- Less affected by outliers.
- **Less Affected** by extreme values in an asymmetric (skewed) distribution.

Most preferred measure of central tendency–MEDIAN

LIMITATIONS:

- Can't be used for nominal data.

MODE:

- It is the observation or value that occurs most frequently in a data set.
- **Most repeated value** (most frequent value).
- Data is arranged from smallest to largest and value which occurs **most often** is Mode.
- **Least affected** measure of central tendency.
- ***Peak of frequency distribution curve or tip of highest bar in histogram is MODE.***

MODE = 3 Median – 2 Mean

OR

MODE = Mean – 3(Mean – Median)

ADVANTAGES:

- Can be used for both numerical and categorical data.
- Very useful for categorical data.

LIMITATIONS:

- If all values are different and data is continuous then mode is least useful.

MUST KNOW POINTS

- **Mean = Average Value; Most Affected**
- **Median = Middle Most Value; Most Preferred**
- **Mode = Most Repeated Value; Least Affected**

Measure of Central Tendency	Appropriate Data Type(s)	Sensitivity to Outliers	Method to Determine
Mean	Normally distributed continuous data sets	Yes	Sum of observations/No. of observations
Median	Asymmetrical continuous data sets Ordinal data sets	No	Middle value within ordered data sets
Mode	Nominal data sets Ordinal data sets	No	Most commonly occurring data point

DEPENDING ON SCALE

Scale	Best Measure of Central Tendency
Nominal	• Mode
Ordinal	• Median
Interval	• Symmetrical distribution = Mean • Skewed distribution = Median
Ratio	• Symmetrical distribution = Mean • Skewed distribution = Median

MULTIPLE CHOICE QUESTIONS

(MEASURES OF CENTRAL TENDENCY)

1. Out of 15 births in a hospital, 7 babies weighed over 3 kg and 7 weighed less than 3 kg. What value do 3 kg represent? [NEET Pattern 2001]

(a) Geometric average
(b) Arithmetic average
(c) Median
(d) Mode

Ans: (c)

- MEDIAN is middle value of data when it is arranged in ascending/descending order.
- When arranged in ascending or descending order, 3 kg will be central value.
- **So 3 kg represents Median.**

2. Number of malaria cases reported during the last 10 years in a town is given below, 250, 320, 190, 300, 5000, 100, 260, 350, 320 and 160. The epidemiologist wants to find out the average number of malaria cases reported in that town during the last 10 years. The most appropriate measure of average for this data will be: [AIIMS May 2014, AIIMS Nov 2004]

(a) Arithmetic mean
(b) Mode
(c) Median
(d) Geometric mean

Ans: (c)

- Since there are extreme values in this data.
- **Median is best measure of central tendency for this data.**
- Mean and mode are affected by extreme values.

3. **Incidence of malaria in an area is 20, 20, 50, 56, 60, 5000, 678, 898, 345, 456. Which of these methods is the best to calculate average incidence? [AIIMS 2012, 2000]**

(a) Arithmetic mean (b) Geometric mean

(c) Median (d) Mode

Ans: (c)

- Since there are extreme values in this data.
- **Median is best measure of central tendency for this data.**
- Mean and mode are affected by extreme values.

4. **In a bimodal series, if mean is 2 and median is 3, what is the mode? [AIIMS 2009]**

(a) 5 (b) 2.5

(c) 4 (d) 3

Ans: (a)

- For bimodal series, Mode = 3 Median – 2 Mean.
- Therefore $(3 \times 3) - (2 \times 2) = 9 - 4$.
- **It equals to 5.**

5. **Mean value of weight of a group of 10 boys was found to be 18.2 kg. Later it was found that weight of one of the boys was wrongly recorded as 2.0 kg that should have been 20 kg. The true mean weight of the group is: [NEET Pattern 2018]**

(a) 18.2 kg (b) 20.2 kg

(c) 16.4 kg (d) 20 kg

Ans: (d)

- Mean is sum of all values divided by number of values.
- Sum of all values is (mean) × (number of values) = $18.2 \times 10 = 182$.
- But it is wrongly interpreted so corrected sum of all values = $182 + (20 - 2) = 200$.
- **True mean= 200/10= 20 kg.**

6. **If 60 values are arranged in ascending order, middle value is: [AIIMS 2009]**

(a) Arithmetic Mean (b) Median

(c) Harmonic Mean (d) Mode

Ans: (b)

- MEDIAN is middle value of data when it is arranged in ascending/descending order.

7. **Median is important for all *except*: [NEET Pattern 2015]**

(a) Blood pressure (b) Survival time

(c) Incubation period (d) Health expenses

Ans: (a)

- For ordinal data best measure is Median.
- Since blood pressure normal scale is used. **Best measure is mean.**

Scale	Best Measure of Central Tendency
Nominal	• Mode
Ordinal	• Median
Interval	• Symmetrical distribution = Mean • Skewed distribution = Median
Ratio	• Symmetrical distribution = Mean • Skewed distribution = Median

8. Values are arranged in ascending and descending order to calculate: **[AIIMS 2000]**

(a) Mean (b) Mode

(c) Median (d) Standard distribution

Ans: (c)

- MEDIAN is middle value of data when it is arranged in ascending/descending order.

9. Best index to detect deviation is:

(a) Range (b) Standard deviation

(c) Mean deviation (d) Variance

Ans: (b)

- Best index to detect deviation is Standard deviation.

10. Median of the following datas will be: 10, 9, 8, 8, 7: **[NEET Pattern 2000]**

(a) 8.75 (b) 8

(c) 9 (d) 10

Ans: (b)

- Data is arranged in descending order.
- Middle most value is 8 which is median.

11. Most frequently occurring value in a group of data: **[NEET Pattern 2012]**

(a) Mean (b) Mode

(c) Median (d) Standard deviation

Ans: (b)

- Most frequently occurring value in data is defined as Mode.

12. True among for given data 20, 31, 31, 31, 25, 28, 35, 38, 31: **[NEET Pattern 2014]**

(a) Mean is 31 (b) Range is 18

(c) Median is 15 (d) Mode is 15

Ans: (b)

- Mean = 270/9 = 30
- Median = 31
- Mode = 31
- **Range = 18 (Maximum value – Minimum value).**

13. Commonly used measures of central tendency are all *except*: **[AIIMS 2005]**

(a) Mean (b) Median
(c) Mode (d) Chi-square test

Ans: (d)

- MEAN, MEDIAN and MODE are measures of central tendency.

14. Mode definition: **[NEET SS Pattern]**

(a) Most frequent value in a given data
(b) Mid value when data is arranged in ascending/descending order
(c) Sum of all the values divided by the number of values in the data
(d) Difference between the largest and smallest value in a data

Ans: (a)

- Most frequent value in a given data/Most commonly occurring value–Mode.

15. Central tendency is given by all *except*: **[NEET Pattern 2013]**

(a) Mean (b) Median
(c) Mode (d) Variance

Ans: (d)

- No need of explanation.

16. In a survey of weight of 10 people, highest sample of 42 kg was entered by mistake as 90 kg. This will affect mean and median as: **[NEET Pattern 2016]**

(a) Mean increases and Median remains same
(b) Median increases and Mean remains same
(c) Mean increases and median also increases
(d) None of the above

Ans: (a)

- Mean increases as it is influenced by outlier. But median is not affected.

17. Which of the following value is greater in law of means:

(a) Arithmetic mean (b) Harmonic mean
(c) Geometric mean (d) None of the above

Ans: (a)

- According to law of means, Arithmetic mean > Harmonic mean > Geometric mean.

CHAPTER 7

MEASURES OF DISPERSION/ VARIABILITY

- A measure of dispersion or variability describes the spread or scattering of the individual values around the central value in a data.
- Variability is the extent values are spread above and below the middle values.

MEASURES OF DISPERSION	
• Range	• Standard error of mean
• Mean or Average Deviation	• Standard error of difference between 2 means
• Standard Deviation	• Standard error of proportion
• Inter-quartile range	• Standard error of difference between 2 proportions
• Coefficient of variation	• Standard error of correlation coefficient
• Variance	• Standard deviation of regression coefficient

RANGE:

- Range is the simplest measure of dispersion.
- It is defined as difference between the highest and lowest figures in a given sample.

EXAMPLE:
Random blood sugar levels of 10 employees in an office is as follows: 112, 105, 210, 98, 116, 129, 156, 174, 88, 133. In this data, highest value is 210 and the lowest is 88. Range is expressed as 88 to 210. **Range = (Maximum Value – Minimum value) = 210 – 88 = 122**

- In case of grouped data, the range is taken as the difference between mid-points of extreme categories.
- The range is not of much practical importance, because it indicates only the extreme values in a data and nothing about dispersion of values between 2 extreme values.
- It gives only a rough idea of the dispersion of the data.
- It is easy to calculate but it is affected by the fluctuations of sampling.

MEAN DEVIATION:

- It is the average of the deviations from the arithmetic mean.
- Mean deviation is the mean of the difference of individual observations from mean, median, mode or any constant observation from the data.

$$MD = \frac{\Sigma(x - \bar{x})}{n}$$

- **'x'** is the value of individual observations; 'X' may be mean, median, mode or constant; n is the total number of observations; and, $\sum$ = is a sign indicating "sum".
- Main drawback of 'Mean Deviation' is that it ignores the algebraic signs. To overcome this drawback another measure of variability called as Variance is calculated.

- Variance is the average of the squared deviations of each of individual value from the mean (*x*).
- It is given by the formula:

$$v = \sqrt{\frac{\Sigma(x - \bar{x})^2}{n - 1}} = SD^2$$

- **<u>When the sample size is less than 30, the denominator (n-1) is used instead of n in both Mean Deviation and Variance calculation.</u>**

STANDARD DEVIATION (SD OR σ)

- Standard deviation is the best and most frequently used measure of dispersion.
- Most appropriate measure of dispersion.
- In simple terms, it is defined as **"Root - Means - Square - Deviation".**
- It is denoted by the Greek letter sigma **'σ'** or by the initials **SD.**
- It gives an idea of the 'spread' of the dispersion, i.e., the larger the standard deviation, the greater the dispersion of values around the mean.
- Used in finding suitable size of sample for valid conclusions.

- When the sample size is more than 30, standard deviation is calculated using formula:

$$\sigma = \sqrt{\frac{\Sigma(x - \bar{X})^2}{n}}$$

- When the sample size is less than 30, denominator (n-1) is used instead of n.

$$\sigma = \sqrt{\frac{\Sigma(x - \bar{X})^2}{n - 1}}$$

- **SD = Square root of Variance** (or) **Variance = (SD)^2**

- With reference to a normal curve 68%, 95% and 99% of the values are observed in X ± 1 SD, X ± 2 SD and X ± 3 SD area respectively.

INTERQUARTILE RANGE:

- Quartiles divide the total number of observations in a data set into 4 equal parts of 25% each.
- Thus there are three quartiles (Q_1, Q_2 and Q_3) which divide the total observations into four equal parts.
- Second quartile Q_2 is equivalent to the middle value, i.e., median.

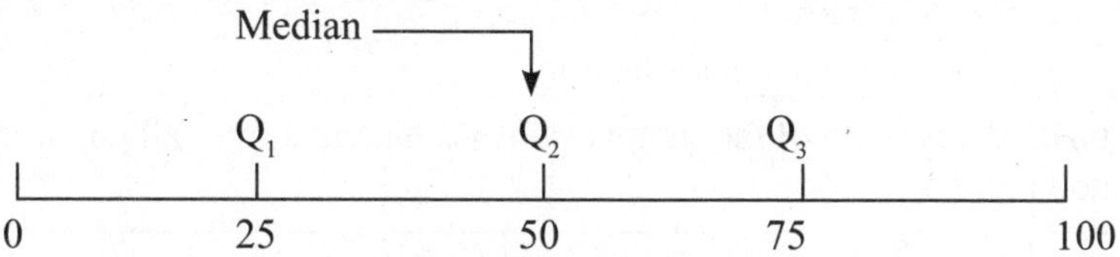

- Interquartile range is the difference in the values of the two extreme quartiles.

Interquartile range = $Q_3 - Q_1$

- Interquartile range gives the middle 50% values of the data set.
- Interquartile range is easy to calculate but it suffers from same defects as that of range.

COEFFICIENT OF VARIATION (CV):

- Coefficient of variation compares the variability in two data sets.
- ***Relative measure of dispersion/variability.***

It measures the variability relative to the mean and is calculated as follows:

$$\text{Coefficient of Variation (CV)} = \frac{\text{Standard Deviation}}{\text{Mean}} \times 100$$

- If the coefficient of variation is greater for one data set, it suggests that the data set is more variable than the other data set.

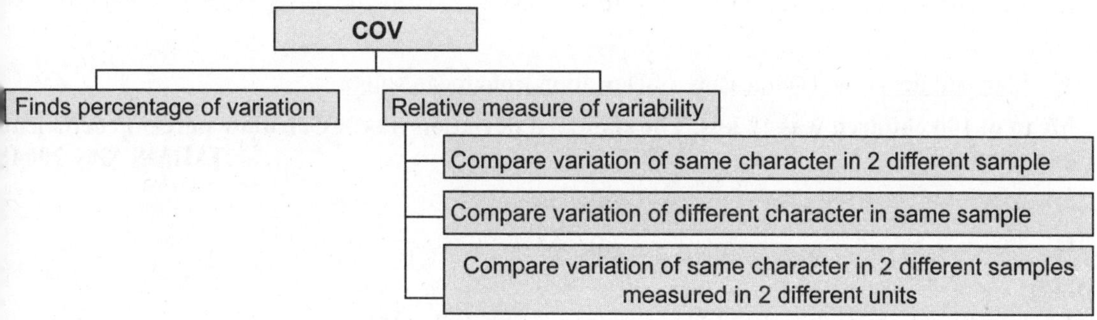

STANDARD ERROR OF MEAN (SE MEAN):

- SE mean is the measure of difference between sample and population values.
- **SE mean = Standard deviation (SD)/ $\sqrt{\text{sample size}}$.**

- Difference between the sample values and the original population values is called Sampling Error. Standard Error is used to estimate this sampling error.

USES OF STANDARD ERROR OF MEAN IN LARGE SAMPLES
• To work out limits of desired confidence within which population mean will lie
• To determine if sample is drawn from a known population or not
• To find SE of difference between 2 means
• To calculate sample size (within desired confidence limits)

- Standard Error is the standard deviation of the sample statistics obtained from all possible samples drawn from the target population.

• Standard Error of Mean = $\sqrt{\frac{\sigma}{n}}$
• Standard Error of Proportion = $\sqrt{pq/n}$
• Standard Error of Difference between Means = $\sqrt{\sigma^2 1/n1 + \sigma^2 2/n2]}$
• Standard Error of Difference between Proportions = $\sqrt{(p1.q1/n1) + (p2.q2/n2)}$

IMPORTANCE OF STANDARD ERROR OF MEAN
• Greater the standard deviation, greater will be standard error
• SE can be minimized by reducing SD and by increasing the sample size

MULTIPLE CHOICE QUESTIONS

(MEASURES OF DISPERSION)

1. Standard deviation is the measure of: **[NEET Pattern 2008]**

(a) Chance (b) Central tendency
(c) Deviation from mean value (d) None

Ans: (c)

- Standard deviation is a measure of Deviation from mean value.

2. Mean of 100 children was 12 kgs. The standard deviation was 3. Calculate percent coefficient of variance: **[AIIMS May 2004]**

(a) 25% (b) 35%
(c) 45% (d) 55%

Ans: (a)

- Coefficient of variance = SD/Mean × 100= 3/12 × 100 = 25%

3. While calculating the incubation period for measles in a group of 25 kids, standard deviation is 2 and mean incubation period is 8 days, calculate standard error: **[NEET SS Pattern]**

(a) 0.4 (b) 1.0
(c) 2.0 (d) 0.5

Ans: (a)

- Standard error is standard deviation divided by square root of n = 2/5 = 0.4.

4. Standard deviation of means measures: **[AIIMS May 2001]**

(a) Non-sampling errors (b) Sampling errors

(c) Random errors (d) Conceptual errors

Ans: (b)

- Standard deviation of means measures sampling errors.
- Sampling errors are not errors in conventional sense.

5. If the birth weight of each of the 10 babies born in a hospital in a day is found to be 2.8 kg, then the standard deviation of this sample will be:

[NEET Pattern 2013, AIIMS May 2006, NEET Pattern 2001]

(a) 2.8 (b) 0

(c) 1 (d) 0.28

Ans: (b)

- Birth weight of each of the 10 babies born in a hospital in a day is found to be 2.8 kg.
- Thus standard deviation is equal to zero.

6. Among 100 women with average Hb of 10 gm%, standard deviation was 1, what is the standard error? **[AIIMS May 01, 04, 07]**

(a) 0.01 (b) 0.1

(c) 1 (d) 10

Ans: (b)

- SE = SD/square root of n = 1/10 = 0.1.

7. If each value of a given group of observations is multiplied by 10, the standard deviation of the resulting observations is: **[NEET Pattern 2004]**

(a) Original standard deviation × 10 (b) Original standard deviation/l0

(c) Original standard deviation – 10 (d) Original standard deviation itself

Ans: (a)

- Since all observations are multiplied by 10 the standard deviation is also multiplied by 10.

8. The Hb level in healthy women has mean 13.5 g/dl and standard deviation 1.5 g/dl, what is the Z score for a woman with Hb level 15.0 g/dl: **[NEET Pattern 2005]**

(a) 9.0 (b) 10.0

(c) 2.0 (d) 1.0

Ans: (d)

- **Z score = Individual value – Mean/SD.**
- Z score = 15 – 13.5/1.5 = 1.

9. In a sample of 100 pregnant females, Mean haemoglobin level estimated was 10 gm% with a standard deviation of 1 gm%. What is the standard error? [NEET Pattern 2012]

(a) 1 gm% (b) 10 gm%

(c) 0.1 gm% (d) 100 gm%

Ans: (c)

- Standard error is standard deviation divided by square root of n = 1/10 = 0.1.

10. Maximum and minimum value is given by: [NEET SS Pattern 2016]

(a) Standard deviation (b) Range

(c) Mean deviation (d) Coefficient of variation

Ans: (b)

- Maximum and minimum values are given by Range.

11. Most common deviation used in social medicine is: [NEET Pattern 2012]

(a) Mean (b) Range

(c) Variance (d) Standard deviation

Ans: (d)

- Most commonly used deviation is standard deviation.

12. In a population of 100 prevalence of candida glabrata was found to be 80%. If the investigator has to repeat prevalence with 95% confidence what will prevalence be? [AIIMS May 2013]

(a) 78-82% (b) 76-84%

(c) 72-88% (d) 74-86%

Ans: (b)

- SE = Square root of pq/n = $\sqrt{pq/n}$
- $\sqrt{pq/n} = \sqrt{p(1-p)/n}$ **(SINCE q = 1 – p)**
- Therefore SE = $\sqrt{0.16/100}$ = 0.4
- Confidence interval is p value +/– 2 Standard error = 76 to 84%.

13. How much population falls between median and median plus one standard deviation in a normal distribution? [AIIMS November 2013]

(a) 0.34 (b) 0.68

(c) 0.17 (d) 0.47

Ans: (a)

- Normal distribution has half of values on each side–Bilaterally symmetrical.
- Median, mean and mode lies at centre.
- Mean + 1 SD cover 68% values (Mean + 1 SD cover 34% values).
- Therefore 34% values lies on plus side and 34% on other side.

14. There is a population of 20000 people with mean hemoglobin being 13.5 gm% having a normal distribution. What proportion of population constitutes proportion more than 13.5 gm%? [AIIMS November 2013]

(a) 0.25 (b) 0.50

(c) 1 (d) 0.34

Ans: (b)

- In normal distribution curve values are arranged in bilaterally symmetrical manner.
- So 50% of values lie on each side of curve.
- 0.50 is the answer.

15. Measuring variation between two different units is done through: [AIIMS November 2014]

(a) Variance (b) Coefficient of variation

(c) Standard deviation (d) Range

Ans: (b)

- Coefficient of variation compares the variability in two data sets.
- If the coefficient of variation is greater for one data set, it suggests that the data set is more variable than the other data set.
- It measures the variability relative to the mean and is calculated as follows: **Coefficient of Variation (CV) = (Standard Deviation/Mean) × 100.**

16. True about standard deviation is/are: [PGI November 2014]

(a) 1 SD covers 95% values

(b) Indicated distribution of variables

(c) Most common method used for dispersion

(d) Better indicator of variance than range

(e) Should be used only in normal distributions

Ans: (b), (c), (d)

- **Refer to notes on standard deviation.**

17. Following denotes measures of variability: [PGI November 2011]

(a) Range (b) Mean deviation

(c) Standard deviation (d) Median

(e) Mode

Ans: (a), (b), (c)

- Question is about measures of dispersion.
- Options A, B, C are measures of dispersion.

18. If variance is 9, SD is: **[NEET Pattern 2017]**

(a) 81 (b) 9

(c) 27 (d) 3

Ans: (d)

- SD = Square root of variance.
- Therefore square root of 9 is 3.
- SD = 3.

19. Correct relation between variance and SD is: **[NEET Pattern 2015]**

(a) SD = Variance square (b) SD = Variance

(c) SD = Square root of Variance (d) Variance = Square root of SD

Ans: (c)

- SD = Square root of Variance.

20. When calculating incubation period for measles in group of 25 children. SD is 2, Mean incubation period is 8 days. Calculate Standard error:

(a) 0.4 (b) 1

(c) 2 (d) 0.5

Ans: (a)

- SE = SD/Square root of n.
- SE = 2/5 = 0.4.

CHAPTER 8

MEASURES OF LOCATION

CENTILE (PERCENTILE)

- Divides a distribution into 100 equal parts.
- Requires 99 intercepts for division into 100 parts.
- **Total percentiles: 99.**
- **Each percentile has 1% observations.**

N^{TH} PERCENTILE IMPLIES
When all values are arranged in ascending order, n% are below this value and (100 – n)% are greater than that number.

METHODS FOR LOCATION OF PERCENTILES:

- **Graphical method:** Cumulative frequency diagram–OGIVE (CFD).
- **Arithmetic method:** Cumulative frequency table.

FORMULA
Pa = (a × N)/100 *Pa is percentile, a is percentile we find, N is no. of subjects*

APPLICATIONS AND USES OF PERCENTILES:

- Location of a percentile.
- Calculate median value of a variable.
- Preparation of a standard percentile (Q2, Median) for particular age, sex, etc.
- Comparison of a percentile value of a variable (between samples or populations).
- Measure of dispersion (interquartile/semi-interquartile range).

Percentiles are positional measures for a distribution

PERCENTILE RANKS:

- Conveys individual rank easily.
- It tells us percentage of scores that usually fall below a given score.

TERTILE

- Divides a distribution into 3 equal parts, so number of intercepts required will be 2.
- T1, T2–**Intercepts.**

> - **T1 (1st tertile)** divides a distribution in a ratio of **1 : 2** (33 : 66)
> - **T2 (2nd tertile)** divides a distribution in a ratio of **2 : 1** (66 : 33)

QUARTILES

- Divides a distribution into 4 equal parts, so the number of intercepts required will be 3, i.e., Q1, Q2, Q3.

> - **Q1 (1st Quartile)** divides a distribution in a ratio of **1 : 3**
> - **Q2 (2nd Quartile)** divides a distribution in a ratio of **1 : 1**
> - **Q3 (3rd Quartile)** divides a distribution in a ratio of **3 : 1**

- If 100% is broken into 4 equal parts subdivisions include 25%, 50% and 75%.

Q1	Q2	Q3	Q4
25%	25%	25%	25%
QR = Q3 – Q1			

INFERENCES

- 1st Quartile (Lower quartile) is at 25th percentile
- ***2nd Quartile (Median) is at 50th percentile***
- 3rd Quartile (Upper quartile) is at 75th percentile

ADVANTAGES: Quartiles are less affected by outliers. So used in many situations as a measure of location.

INTERQUARTILE RANGE

- Difference between Upper and Lower quartiles
- Q3-Q1
- This range is middle or 50% of data

Semiquartile range is half of quartile range = Q3-Q1/2

PENTILE (QUINTILE)

- Divides a distribution into 5 equal parts, so number of intercepts required will be 4.
- P1, P2, P3, P4–**Intercepts.**

- **P1 (1st pentile)** divides a distribution in a ratio of **1 : 4** (20 : 80)
- **P2 (2nd pentile)** divides a distribution in a ratio of **2 : 3** (40 : 60)
- **P3 (3rd pentile)** divides a distribution in a ratio of **3 : 2** (60 : 40)
- **P4 (4th pentile)** divides a distribution in a ratio of **4 : 1** (80 : 20)

	Divides distribution into	**No. of intercepts**
Tertile	3 equal parts	2
Quartile	4 equal parts	3
Pentile (Quintile)	5 equal parts	4
Hextile	6 equal parts	5
Heptile	7 equal parts	6
Octile	8 equal parts	7
Decile	10 equal parts	9
Centile (Percentile)	100 equal parts	99

MULTIPLE CHOICE QUESTIONS

1. 'Centile' divides data into: **[NEET Pattern 2010]**

(a) 100 equal parts (b) 4 equal parts
(c) 10 equal parts (d) 20 equal parts

Ans: (a)

- Centile divides data into 100 equal parts.
- Requires 99 intercepts for division into 100 parts.
- **Total percentiles: 99.**
- **Each percentile has 1% observations.**

2. 50th percentile is equivalent to: **[AIIMS 2007]**

(a) Mean (b) Median
(c) Mode (d) Range

Ans: (b)

INFERENCES

- 1st Quartile (Lower quartile) is at 25th percentile
- ***2nd Quartile (Median) is at 50th percentile***
- 3rd Quartile (Upper quartile) is at 75th percentile

3. A bacterium can divide every 20 minutes. Beginning with single individual, how many bacteria will be there in population if there is exponential growth for 3 hours? **[AIIMS 2005]**

(a) 18 (b) 440
(c) 512 (d) 1024

Ans: (c)

- No formulas for this question.
- Solve this by writing no of 20 min in 3 hours (Approximately 9 divisions).
- For every 20 min one becomes two.
- Or simply it can be expressed as 2^n where n is no. of divisions.
- Therefore $2^9 = 512$.

4. For a group of n = 250 subjects, 40th percentile would be the following value (after arranging in ascending order): [NEET Pattern 2015]

(a) 7th (b) 40th

(c) 100th (d) 140th

Ans: (c)

- 40th percentile implies 40 × 250/100 = 100.

CALCULATE
Pn = (n × N)/100
Pn is percentile, n is percentile we find, N is no.of subjects

5. Central value of a set of 180 values can be obtained by: [AIIMS 2010]

(a) 90th percentile (b) 2nd tertile

(d) 9th decile (d) 2nd quartile

Ans: (d)

- 2nd Quartile is defined as median.
- This is the central value.

INFERENCES
• 1st Quartile (Lower quartile) is at 25th percentile
• ***2nd Quartile (Median) is at 50th percentile***
• 3rd Quartile (Upper quartile) is at 75th percentile

6. Measure of location that divides distribution in ratio of 3:1: [NEET Pattern 2013]

(a) Median (b) First quartile

(c) Third quartile (d) Mode

Ans: (c)

- **Q1 (1st Quartile)** divides a distribution in a ratio of **1 : 3.**
- **Q2 (2nd Quartile)** divides a distribution in a ratio of **1 : 1.**
- **Q3 (3rd Quartile)** divides a distribution in a ratio of **3 : 1.**

7. **Chest physician observed that FEV distribution in 300 smokers with median of 2.5 litres with 1st quartile–1.5 litres and 3rd quartile–3.5 litres. Based on this data how many persons in sample are expected to have FEV between 1.5 to 3.5 litres:**

(a) 75 (b) 150

(c) 300 (d) 250

Ans: (d)

DIVIDE DATA INTO 4 QUARTILE

- 1st quartile–75
- 2nd quartile–150
- 3rd quartile–225
- 4th quartile–300

So, numbers between 1st and 3rd quartile = 225 – 75 = 150

8. **Length of box in Box whisker plot is:**

(a) Median (b) Interquartile range

(c) Whiskers (d) Any of the these

Ans: (b)

- Used to present Quantitative Data.
- Measured in Quartiles.
- Helps in General assessment of trend of data with respect to higher or lower values.
- Comparison of ≥ 2 Quantitative Variables can be done.
- Minimum is lower quartile.
- Maximum is upper quartile.
- Box is rectangle which encloses middle half of sample.
- Length of box is interquartile range.
- Whiskers sprout from 2 ends of box until they reach sample maximum and minimum values.

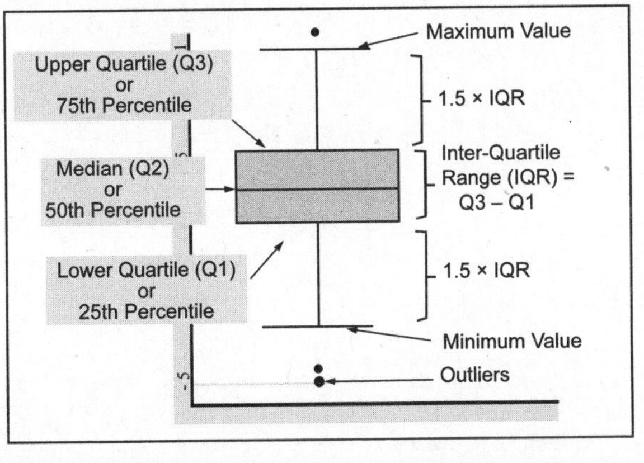

9. **Blood pressure of data of 200 persons were collected. First quartile is 90 mm Hg and Third quartile is 115 mm Hg. How many patients will have BP between 3rd and 4th quartile?**

(a) 25 (b) 50

(c) 100 (d) 200

Ans: (b)

DIVIDE DATA INTO 4 QUARTILE

- 1st quartile–50
- 2nd quartile–100
- 3rd quartile–150
- 4th quartile–200

So, numbers between 1st and 3rd quartile = 200 – 150 = 50

10. **Measure of location that divides distribution to 1:3 ratio is:**

(a) Median (b) 1st quartile

(c) Third quartile (d) Mode

Ans: (b)

- **Q1 (1st Quartile)** divides a distribution in a ratio of **1 : 3.**
- **Q2 (2nd Quartile)** divides a distribution in a ratio of **1 : 1.**
- **Q3 (3rd Quartile)** divides a distribution in a ratio of **3 : 1.**

11. **Which of the following is not a positional measure?**

(a) Median (b) Tertile

(c) Quartile (d) Mean

Ans: (d)

- Mean is a measure of central tendency not positional measure.
- **Median represents 2nd quartile, so it can be placed as positional measure.**
- Quartile and tertile are obviously Positional measures.

CHAPTER 9

HYPOTHESIS TESTING AND CONCEPT OF p VALUE

HYPOTHESIS TESTING:

- It is done to find evidence for or against a hypothesis assumed about the population, based on sample data.

OBJECTIVES OF HYPOTHESIS TESTING
• To find out if the data from the sample is same as entire population or not.
• To find out if any association exists between ≥ 2 variables in a data set and whether it is statistically significant or not.
• To find out if any difference exists between ≥ 2 variables in a data set and whether it is statistically significant or not.

NULL HYPOTHESIS:

- It assumes no significant difference between 2 samples, whatever difference that occurs is by chance.

ALTERNATE HYPOTHESIS:

- There is significant difference between 2 samples.

p VALUE:

- 'p' value is the actual probability of committing type 1 error.
- If $p < \alpha$ [usually taken as 5% or 0.05], association is statistically significant and null hypothesis is rejected.
- Smaller the p Value stronger the evidence against null hypothesis and in favour of alternate hypothesis.

Research Study Result	**Reality: Null Hypothesis (H_0)**	
	True	*False*
Accept	Correct	β error (Type II error)
Reject	α error (Type I error)	Correct

TYPE I ERROR OR α ERROR
• Rejecting a null hypothesis when actually it is true (statistically significant difference is found between two groups when actually there is no difference).
• If null hypothesis is false there can be no type I error.

Contd.

Contd.

- Probability of a type I error is equal to alpha.
- Value of alpha is called as significance level.
- Multiple comparisons and repeated testing for significance increase likelihood of type I error.
- **APPLICATION–Sample size determination and Statistical testing is based on α error (α = 0.05).**
- *When hypothesis testing results in p value less than significance level result of hypothesis test is statistically significant.*

TYPE 2 ERROR OR β ERROR

- Acceptance of a null hypothesis as true when actually it is false.
- Study fails to find a statistically significant difference when actually there is a difference.
- If null hypothesis is true there can be no type II error.
- Probability of a type II error is equal to beta.
- It is inversely related to type I error.
- **APPLICATION–It is used to determine Power of study = 1 – β**

POWER OF STUDY

- It is probability that study would reject null hypothesis as false when it is actually false.
- Ability of study to find significant difference between if it is present.
- Conventionally a study is required to have power of 0.8 to be acceptable, i.e., study that has less than 80% chance of detecting a false null hypothesis is unacceptable.
- **Power is proportional to sample size.**
- *So, power of study is increased by increase in sample size, reducing type II error and increased sensitivity.*
- Power of statistical test is a numeric representation of sensitivity.

Power of study = (1 – β)

CONFIDENCE LEVEL (1 - α):

- It is the probability that a sample statistic or a difference detected between the 2 samples is the same as in real population, majority of the times and occurrence of an error due to chance is very less.
- Is the 'degree of assurance for an interval to contain the value of the parameter'?
- 100% confidence level can never be achieved.
- Internationally accepted confidence level is 95%.
- **So alpha value is fixed at 5% and p value is most commonly < 0.05.**

Type of Error	Synonym	Explanation	Conventional Value
α	Type I error/P value/Significance level/ False Positive error/Allowable error	Probability of rejecting the null hypothesis when it is true in reality	0.05 or 0.01
β	Type II error / False Negative error	Probability of accepting the null hypothesis when it is in reality	0.20 or 0.10

Contd.

Contd.

$1 - \alpha$	Correct Decision / Confidence level	Probability of accepting the null hypothesis when it is true in reality	0.95 or 0.99
$1 - \beta$	Correct Decision/Power of the study	Probability of rejecting the null hypothesis when it is false in reality	0.80 or 0.90

Alpha value fixed at 5%	
p Value < 5%	p Value > 5%
Reject null hypothesis Accept study Significant difference between treatment and placebo	Accept null hypothesis Reject study No Significant difference between treatment and placebo

MULTIPLE CHOICE QUESTIONS

1. Type I error is classified as: **[NEET Pattern 2010]**

(a) Alpha error (b) Beta error

(c) Gamma error (d) Delta error

Ans: (a)

TYPE I ERROR OR α ERROR

- Rejecting a null hypothesis when actually it is true (statistically significant difference is found between two groups when actually there is no difference).
- If null hypothesis is false there can be no type I error.
- Probability of a type I error is equal to alpha.
- Value of alpha is called as significance level.

2. "p" value of a randomized controlled trial comparing Operation A (new procedure) and Operation B (Gold standard) is 0.04. From this, we conclude that:

(a) Type II error is small and we can accept the findings of study

(b) Probability of false negative conclusion that Operation A is better than Operation B, when in truth it is not, is 4%

(c) Power of study to detect a difference between Operation A and B is 96%

(d) Probability of false positive conclusion that 'Operation A is better that Operation B', when in truth it is not, is 4%

Ans: (d)

- "p" value of a randomized controlled trial comparing Operation A (new procedure) and Operation B (Gold standard) is 0.04.
- Thus probability of committing Type I error is 0.04 or 4% (i.e., probability of declaring a false positive conclusion that 'Operation A is better that Operation B', when in truth it is not, is 4%).

3. Power of study can be increased by: **[AIIMS 2014, NEET Pattern 2012]**

(a) Increasing alpha error (b) Decreasing beta error

(c) Decreasing alpha error (d) Increasing beta error

Ans: (b)

- Power of study is increased by increase in sample size, reducing type II error and increased sensitivity.

4. In assessing association between maternal nutritional status and birth weight of newborns, two investigators A and B studied separately and found significant results with p values 0.02 and 0.04 respectively. From this information, what can you infer about magnitudes of association found by 2 investigations? **[AIIMS 2014]**

(a) The magnitude of association found by investigator A is more than that found by B

(b) The magnitude of association found by investigator B is more than that found by A

(c) The estimates of association obtained by A and B will be equal, since both are significant

(d) Nothing can be concluded as the information given is inadequate

Ans: (d)

- Assessment of association between maternal nutritional status and birth weight of newborns, two investigators A and B studied separately and found significant results with p values 0.02 and 0.04 respectively.
- *A has 98% chance of being correct whereas B has 96% chance of being correct.*
- Only p values are given we can't decide magnitude of association with this data as both values are significant but confidence levels are not mentioned.

5. A randomized trial comparing efficacy of two regimens showed that difference is statistically significant with p<0.001 but in reality two drugs do not differ in their efficacy. This is an example of: **[AIIMS 2016]**

(a) Type I error (b) Type II error

(c) 1-alpha (d) 1-beta

Ans: (a)

- When hypothesis testing results in p value less than significance level result of hypothesis test is statistically significant.
- Then it is described as type 1 error.

6. After applying a statistical test, an investigator gets the 'P value' as 0.01. It means that: **[AIIMS 2003, 2005, 2008]**

(a) The probability of finding a significant difference is 1%

(b) The probability of declaring a significant difference is 1%

(c) The difference is not significant 1% times and significant 99% times

(d) The power of the test used is 99%

Ans: (b)

- 'p' value is the actual probability of committing type 1 error.
- If, $p < \alpha$ [usually taken as 5% or 0.05], association is statistically significant and null hypothesis is rejected.

7. All are true about P-value *except*: **[AIPGME 03]**

(a) It is the probability of committing Type I error

(b) It is equal to 1-beta

(c) It is chance that the presence of difference is concluded when actually there is none

(d) When P-value is less than α, the result is statistically significant

Ans: (b)

- Power of the study is equal to 1-beta.

8. All are true *except*: **[AIIMS 2004]**

(a) Alpha is the maximum tolerable probability of type I error

(b) Beta is the probability of type II error

(c) When null hypothesis is true but is rejected, it is type II error

(d) P-value can be more or less than alpha

Ans: (c)

Research Study Result	Reality: Null Hypothesis (H_0)	
	True	*False*
Accept	Correct	β error (type II error)
Reject	α error (type I error)	Correct

9. Statistical Power of a trial is equal to: **[AIIMS 2014]**

(a) 1 + alpha (b) 1 – beta

(c) 1 + beta (d) 1 – alpha

Ans: (b)

- Power of study = $(1 - \beta)$.

10. P-value is the probability of: **[AIIMS 2010]**

(a) Not rejecting a null hypothesis when true (b) Rejecting a null hypothesis when true

(c) Not rejecting a null hypothesis when false (d) Rejecting a null hypothesis when false

Ans: (b)

- 'p' value is the actual probability of committing type 1 error.
- If, $p < \alpha$ [usually taken as 5% or 0.05], association is statistically significant and null hypothesis is rejected.

11. A randomized trial comparing the efficacy of two drugs showed a difference between the two

with a p value of <0.005. In reality, however the two drugs do not differ. This is therefore an example of: [AIIMS 2012]

(a) Type I error (b) Type II error

(c) 1 – alpha (d) 1 – beta

Ans: (a)

- A test with a high specificity has a low Type I error rate.

12. p-value is defined as: [NEET Pattern 2014]

(a) Probability of declaring a significant difference when actually it is not present

(b) Probability of declaring a significant difference when actually it is present

(c) Probability of not declaring a significant difference when actually it is not present

(d) Probability of not declaring a significant difference when actually it is present

Ans: (a)

- 'p' value is the actual probability of committing type 1 error.

13. Rejecting a null hypothesis when it is true is called as: [NEET Pattern 2010]

(a) Type 1 error (b) Type 2 error

(c) Type 3 error (d) Type 4 error

Ans: (a)

Type of Error	Synonym	Explanation	Conventional Value
α	Type I error/P value/Significance level/ False Positive error/ Allowable error	Probability of rejecting the null hypothesis when it is true in reality	0.05 or 0.01
β	Type II error / False Negative error	Probability of accepting the null hypothesis when it is in reality	0.20 or 0.10
1 – α	Correct Decision / Confidence level	Probability of accepting the null hypothesis when it is true in reality	0.95 or 0.99
1 – β	Correct Decision/Power of the study	Probability of rejecting the null hypothesis when it is false in reality	0.80 or 0.90

14. When we say that "difference is significant", it means that: [NEET Pattern 2016]

(a) It is likely by chance and when P> 0.0.5 (b) It is unlikely by chance and when P > 0.05

(c) It is unlikely by chance and when P < 0.05 (d) It is likely by chance and when P < 0.05

Ans: (c)

- *When hypothesis testing results in p value less than significance level result of hypothesis test is statistically significant.*
- If, p < α [usually taken as 5% or 0.05], association is statistically significant and null hypothesis is rejected.

15. P value significant indicates: **[NEET Pattern 2013]**

(a) Probability of Type I Error is < 0.05
(b) Probability null hypothesis is correct
(c) Probability null hypothesis is false
(d) To find out meaning of regression

Ans: (a)

- *When hypothesis testing results in p value less than significance level result of hypothesis test is statistically significant.*
- If, $p < \alpha$ [usually taken as 5% or 0.05], association is statistically significant and null hypothesis is rejected.

16. In a test of significance, P value is 0.023 observed difference in study can be considered as: **[NEET Pattern 2015]**

(a) Null hypothesis accepted and the study is rejected
(b) Null hypothesis rejected and the study is accepted
(c) Null hypothesis accepted and the study is accepted
(d) Null hypothesis rejected and the study is also rejected

Ans: (b)

Type of Error	Synonym	Explanation	Conventional Value
α	Type I error/P value/Significance level/ False Positive error/ Allowable error	Probability of rejecting the null hypothesis when it is true in reality	0.05 or 0.01
β	Type II error / False Negative error	Probability of accepting the null hypothesis when it is in reality	0.20 or 0.10
1 – α	Correct Decision / Confidence level	Probability of accepting the null hypothesis when it is true in reality	0.95 or 0.99
1 – β	Correct Decision/Power of the study	Probability of rejecting the null hypothesis when it is false in reality	0.80 or 0.90

17. Significance level is related to:

(a) Type 1 error
(b) Type 2 error
(c) Population mean
(d) Random error

Ans: (a)

- Alpha error is type 1 error or level of significance.
- P Value of 0.05 corresponds to alpha level of 5%.

18. Which of the following is a serious error:

(a) Type 1
(b) Type 2
(c) Type 3
(d) Type 4

Ans: (a)

- Type 1 error is more serious than type 2 error.

19. A researcher conducted a trial with 5% significance level and evaluated use of carnitine containing drug for Alzheimer's disease. Regulatory authority granted clearance for trial on 100 patients as under lab circumstances, drug had shown effect on Alzheimer patients. On further analysis of research it was found that carnitine drug showed an improvement in disease with p value of 0.82. This is an example of:

(a) High alpha error
(b) High beta error
(c) High power of research
(d) High protective power of drug

Ans: (b)

- *Null hypothesis is actually false or wrong but acceptance of null hypothesis by chance then there is beta error.*

20. Value of alpha corresponds to:

(a) Type 1 error
(b) Type 2 error
(c) Selection bias
(d) False –ve results

Ans: (a)

- Alpha error corresponds to type 1 error/p Value.

RECENTLY ASKED TOPIC

EFFECT SIZE

- The **effect size** is the difference between the true value and the value specified in the null hypothesis
- Effect size = True value – Hypothesized value
- For Example → suppose the null hypothesis states that a population mean is equal to 100. A researcher might ask: What is the probability of rejecting the null hypothesis if the true population mean is equal to 90? In this example, the effect size would be 90 – 100, which equals = 10

CHAPTER

10

SAMPLING

SAMPLE:

- Sample is the portion of the population selected by a process called sampling.
- Instead of studying the entire population only a sample is studied.

SAMPLE SIZE DEPENDS UPON
• Effect size (usually the difference between 2 groups) • Population standard deviation (for continuous data) • Desired power of the experiment to detect postulated effect • Significance level (alpha)

SAMPLING FRAME:

- Listing of accessible population from which we draw our sample.

SAMPLING VARIATION:

- When the two samples of same size are drawn from a population their means show variation with reference to each other and with reference to the population mean.
- This variation is called as sampling variation.
- **Depends on two factors:**
  - Variation in the universe (population) and Sample size.

Lesser the sample size more will be the variation

CALCULATION OF SAMPLE SIZE

Sample size = (z × z) × (p × q)/ l × l

- ***"z" is standard deviation*** (Usually standard deviation is taken as 2 SD if z is not given as it is for 95% Confidence limits)
- *"p" is prevalence, "q" is (100 – prevalence)*
- *"l" is allowable error*

SAMPLING:

Probability vs Non-Probability Sampling

Probability Sampling	Non-Probability Sampling
• Each subject has a known/EQUAL probability of being selected • Allows application of statistical sampling theory to results to: – Test hypotheses – Generalise results	• Probability of being chosen in unknown/unequal • Cheaper • Used for Rare cases • Unable to generalise results • Potential for bias is high
Probability Sampling Techniques • Simple Random sampling • Systematic Random sampling • Stratified Random sampling • Cluster sampling • Multi-stage sampling	**Non-Probability Sampling Techniques** • Convenience sampling • Purposive sampling • Snowball sampling • Quota sampling

PROBABILITY SAMPLING

SIMPLE RANDOM SAMPLING:

- **Principle**: Equal chance for each statistical unit
- **Procedure**: Number all units → Randomly draw units
- **Advantages**: Simple, Sampling error easily measured
- **Disadvantages**: Need complete list of units. Does not always achieve best representatively

- Each of the units in sampling frame is assigned a number.
- A table of random numbers lottery method or computer software is used to determine which units are to be included in the sample.
- **Each unit has an equal chance of being drawn in the sample.**
- This technique provides the greatest number of possible samples.
- Applicable when the population is small, homogenous and readily available.
- **It doesn't have observer or selection bias.**

SYSTEMATIC RANDOM SAMPLING

- **Principle:** A unit drawn every k units. Equal chance of being drawn for each unit
- **Procedure:** Calculate sampling interval → Draw a random number (≤ k) for starting → Draw every k units from first unit
- **Advantages:** Ensures representativeness across list. Easy to implement

- No pre numbering is necessary (Sampling frame not required).
- If the sample to be drawn = n, The total population = N.

Then quotient "r" = N/n

- A unit is selected at random and other units are selected by adding "r" to previous selected number or every rth unit at regular intervals is picked.
- Each unit in sampling frame has same chance of being selected, but number of possible samples is greatly reduced.
- Inferior compared to simple random sampling.
- **Preferred when population is large, scattered and not homogeneous.**

STRATIFIED RANDOM SAMPLING:

- **Principle:** Classify population into homogeneous subgroups (strata) → Draw sample in each strata → Combine results of all strata
- **Advantage:** More precise if variable associated with strata. All subgroups represented, allowing separate conclusions about each of them

"It is best done if population is non-homogeneous"

- Sample is drawn in a way that each portion of sample represents corresponding strata of universe.
- Useful where population is heterogenous with regard to characteristic under study/when characteristic is influenced by different sections of population.
- Entire population is divided into homogeneous sub groups/strata depending on characteristic to be studied and simple random sampling is done from each strata.
- Precision of the estimate of the characteristic under study is increased and due representation of population is maintained.

CLUSTER SAMPLING (CRS):

- Applicable when units of population are natural groups or clusters.
- **Clusters are heterogeneous within themselves but homogeneous with respect to each other.**
- Sampling interval is also calculated in CRS.
- Accuracy: Low error rate of only ± 5%.

CLUSTER SAMPLING Vs STRATIFIED SAMPLING

- In stratified sampling random sample is drawn from each of the strata. In Cluster sampling only selected clusters are studied.
- ***Main objective of stratified sampling is increased precision and main objective of cluster sampling is increased efficiency.***
- Analysis in cluster sampling is done on population of clusters whereas in stratified sampling it is done on individual elements within strata.

- **ADVANTAGES:** Simple as no list of units required. Less travel/resources required.

- **DISADVANTAGES:** Clusters cannot be compared with each other. Associated with bias due to intercluster variations.
- Useful for evaluation of programmes and schemes.
- **Example:** Evaluation of immunization coverage.

Best done for homogenous population

Multi Stage Sampling	**Multi Phase Sampling**
• Multistage – large portion of population is under study • Select the population stage by stage • **Example:** Total country	When the study demands more money and time – in order to reduce this–sub sample is taken from a large sample after screening

NON PROBABILITY SAMPLING

Accidental sampling

- Also called as ***Haphazard or convenience sampling***
- Man on street interviews conducted

Expert sampling

- Involves assembly of sample of persons with known expertise

Quota sampling

- Select people randomly according to fixed quota

Heterogeneity sampling

- It is used when we want to include all opinions or views
- Also called as Sampling for diversity

Snowball sampling

- Try to reach population that are inaccessible or hard to find
- Identify drug abuser and ask him about other abusers known to him

VALIDITY

EXTERNAL

Generalization of results back to population

Depends on sampling techniques

Random sampling will improve external validity

INTERNAL

Results should be valid for sample we study

Doesn't depend on sampling techniques

Depends on standard instruments and techniques used in study

MULTIPLE CHOICE QUESTIONS

1. For calculation of sample size for a prevalence study all of the following are necessary *except*: [NEET Pattern 2013]

(a) Prevalence of disease in population (b) Power of the study

(c) Significance level (d) Desired precision

Ans: (d)

SAMPLE SIZE DEPENDS UPON

- Effect size (usually the difference between 2 groups).
- Population standard deviation (for continuous data).
- Desired power of the experiment to detect the postulated effect (Power = 1 – β).
- Significance level.

2. In maternal and child welfare program sampling is done by which method? [NEET Pattern 2010]

(a) Systematic (b) Stratified

(c) Group (d) Cluster-30

Ans: (d)

3. Sampling method used in assessing immunization status of children under immunization program is: [AIIMS 2004, NEET Pattern 2017]

(a) Systematic sampling (b) Stratified sampling

(c) Group sampling (d) Cluster sampling

Ans: (d)

- Applicable when units of population are natural groups or clusters.
- **Clusters are heterogeneous within themselves but homogenous with respect to each other.**
- Sampling interval is also calculated in CRS.
- Accuracy: Low error rate of only ± 5%.

4. Following are sampling techniques used to conduct community health surveys *except*: [AIIMS 2014]

(a) Simple random (b) Systematic random

(c) Stratified random (d) Cluster testing

Ans: (d)

CLUSTER RANDOM SAMPLING

- Adjustment used in few study/sampling designs to allow for design structure, especially to allow for correlations among cluster of observations.

USES:

- Cluster randomised trials.
- Cluster random sampling.
- Multistage sampling.
- Health facility cluster survey.

5. Number of patients required in a clinical trial to treat a specify disease increases as: [NEET Pattern 2015, AIIMS 2012]

(a) Incidence of the disease decreases

(b) Significance level increases

(c) Size of the expected treatment effect increased

(d) Drop-out rate increases

Ans: (d)

6. In a study, people are separated into certain sub-groups and then some are selected randomly from each of these sub-groups. What type of sampling is being done? [NEET Pattern 2014]

(a) Simple random sampling
(b) Cluster random sampling
(c) Systematic random sampling
(d) Stratified random sampling

Ans: (d)

STRATIFIED RANDOM SAMPLING

- **Principle:** Classify population into homogeneous subgroups (strata) → Draw sample in each strata → Combine results of all strata
- **Advantage:** More precise if variable associated with strata. All subgroups represented, allowing separate conclusions about each of them

"It is best done if population is non-homogeneous"

7. True about cluster sampling all *except*: [AIIMS 2011]

(a) Sample size same as simple random
(b) It is two stage sampling
(c) Cheaper than other methods
(d) It is a method for rapid assessment

Ans: (a)

CLUSTER RANDOM SAMPLING

- Sample size is much smaller than that of random sampling.
- It is a 2 stage sampling procedure.
- Cheaper quick method: Lesser time, lesser population coverage, lesser sample size.

8. For which of the following sampling, a Design effect is used? **[NEET Pattern 2013]**

(a) Simple random sampling
(b) Systematic random sampling
(c) Stratified random sampling
(d) Cluster random sampling

Ans: (d)

CLUSTER RANDOM SAMPLING

- Adjustment used in few study/sampling designs to allow for design structure, especially to allow for correlations among cluster of observations.

USES:

- Cluster randomised trials.
- Cluster random sampling.
- Multistage sampling.
- Health facility cluster survey.

9. Population of 50% having disease with estimated prevalence to be 45-55% with 95% of probability of identifying them minimum sample size required is: **[AIIMS 2013]**

(a) 100
(b) 200
(c) 300
(d) 400

Ans: (d)

- **Sample size = (z × z) × (p × q)/ l × l.**
- "z" is standard deviation (Usually standard deviation is taken as 2SD if z is not given as it is for 95% Confidence limits).
- "p" is prevalence, "q" is (100 – prevalence).
- "l" is allowable error.
- p = 50% (50/100); q = 1 – p = 1 – 0.50 = 0.50 (50/100); L = 5% (5/100 as range permissible is 45-50%, i.e., +5%).
- **So, Sample size = [4*50/100*50/100]/ [5/100]2 = 400.**

10. Simple random sampling is ideal for: **[DNB Pattern 2010]**

(a) Vaccinated people
(b) Heterogenous population
(c) Homogenous population
(d) All of the above

Ans: (c)

SIMPLE RANDOM SAMPLING

- Each of the units in sampling frame is assigned a number.
- A table of random numbers lottery method or computer software is used to determine which units are to be included in the sample.

- **Each unit has an equal chance of being drawn in the sample.**
- This technique provides the greatest number of possible samples.
- Applicable when the population is small, homogenous and readily available.
- **It doesn't have observer or selection bias.**

11. Stratified sampling is ideal for: **[NEET Pattern 2012]**

(a) Heterogeneous data (b) Homogenous data
(c) Both (d) None

Ans: (a)

- **Principle:** Classify population into homogeneous subgroups (strata) → Draw sample in each strata → Combine results of all strata
- **Advantage:** More precise if variable associated with strata. All subgroups represented, allowing separate conclusions about each of them

"It is best done if population is non-homogeneous"

12. Children surveyed in cluster sampling for coverage of national immunization programme in: **[DNB Pattern 2011]**

(a) 30 cluster of 5 children (b) 20 cluster of 5 children
(c) 30 cluster of 10 children (d) 30 cluster of 7 children

Ans: (d)

13. In a study first schools are sampled, then sections, and finally students. This type of sampling is known as: **[AIIMS 2012]**

(a) Stratified sampling (b) Simple random sampling
(c) Cluster sampling (d) Multistage sampling

Ans: (d)

Multi Stage Sampling	Multi Phase Sampling
• Multistage – large portion of population is under study • Select the population stage by stage • **Example:** Total country	When the study demands more money and time – in order to reduce this–sub sample is taken from a large sample after screening

14. When number of observations is 25, number of class intervals must be: **[NEET Pattern 2013]**

(a) 25 (b) 15
(c) 10 (d) 5

Ans: (d)

15. All of the following comes under random sampling method *except*: **[DNB Pattern 2011]**

(a) Quota sampling (b) Simple random sampling
(c) Stratified sampling (d) Cluster sampling

Ans: (a)

Probability Vs. Non-Probability Sampling:

Probability Sampling	Non-Probability Sampling
• Each subject has a known/EQUAL of being selected • Allows application of statistical sampling theory to results to: ◆ Test hypotheses ◆ Generalise results	• Probability of being chosen in unknown/unequal • Cheaper • Used for Rare cases • Unable to generalise results • Potential for bias is high
Probability Sampling Techniques • Simple Random sampling • Systematic Random sampling • Stratified Random sampling • Cluster sampling • Multi-stage sampling	**Non-Probability Sampling Techniques** • Convenience sampling • Purposive sampling • Snowball sampling • Quota sampling

16. For an epidemiological study, every 10th person is selected from a population. This type of sampling is known as: **[AIIMS 2014]**

(a) Simple random sampling (b) Stratified random sampling

(c) Systematic random sampling (d) Cluster random sampling

Ans: (c)

SYSTEMATIC RANDOM SAMPLING

- **Principle:** A unit drawn every k units. Equal chance of being drawn for each unit
- **Procedure:** Calculate sampling interval → Draw a random number ($\leq$ k) for starting → Draw every k units from first unit
- **Advantages:** Ensures representativeness across list. Easy to implement.

17. Simple random sampling: **[NEET Pattern 2016]**

(a) Provides least number of possible samples

(b) Equal chance to each for collection of certain number for a sample

(c) Picking every 5th or 10th at regular intervals

(d) Sample represent, a corresponding strata of universe

Ans: (b)

SIMPLE RANDOM SAMPLING

- **Principle**: Equal chance for each statistical unit
- **Procedure**: Number all units → Randomly draw units
- **Advantages**: Simple, Sampling error easily measured
- **Disadvantages**: Need complete list of units. Does not always achieve best representativity

18. Sample is:

(a) All units of population
(b) 50% of population
(c) 10% of population
(d) Any fraction of population

Ans: (d)

- Sample is a part or fraction of population.

19. Calculate sample size of study on Diabetes with prevalence 50% and allowable error 10% at 95% confidence limits:

(a) 100
(b) 200
(c) 400
(d) 1000

Ans: (c)

- **Sample size = (z × z) × (p × q)/ l × l.**
- "z" is standard deviation = 2 (95% Confidence limits).
- "p" is prevalence = 50%, "q" is (100 – prevalence) = 50%, "l" is allowable error = 10%.
- p = 50% q = 1 – p = 50%; L = 10%.
- **So, Sample size = [2 × 2 × 50 × 50]/ [10 × 10] = 400.**

NUTSHELL		
Method	**Reliability**	**Remark**
Simple	Very good	Need full sampling frame Expensive
Systematic	Good	Size of target population needed No need for sampling frame **Economical**
Stratified	Good	Non homogenous population Expensive
Cluster	Poor	Geographically diverse population Very economical
Multistage	Medium	Need sampling frame for each nested unit only

CHAPTER 11

DISTRIBUTION–NORMAL AND SKEWED

DIFFERENCE BETWEEN POPULATION AND SAMPLE:

Population (parameter)	Sample (statistics)
Data from a population: **Mean = μ; SD = σ**	Data from a sample: **Mean = X; SD = s**

NORMAL DISTRIBUTION CURVE (STANDARD)

PROPERTIES OF NORMAL CURVE:

- Smooth Bell Shaped Curve
- Bilaterally Symmetrical Distribution
- Mean = Median = Mode
- Mean = 0; SD = 1; variance = 1
- Area Under the Curve = 1

Ideally, the tails of the curve extends from - ∞ to + ∞, hence does not touches the 'x' axis.

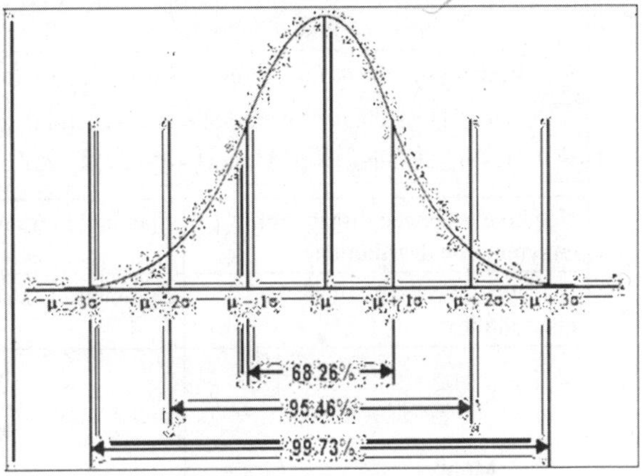

> Mean ± 1 SD covers 68.3% of values, hence called as 68% confidence limit
>
> Mean ± 2 SD covers 95.4% of values, hence called as 95% confidence limit
>
> Mean ± 3 SD covers 99.7% of values, hence called as 99% confidence limit

Z SCORE (STANDARD SCORE):

- **Also known as 'normal deviate'.**
- Standard score indicates how many standard deviations and observation is above or below mean.
- **Z score = 2:** Any measurement of atleast 2SD away is considered too far away to be normal.
- Difference of a value from group mean, in terms of how many times of SD.

> - **Z score** = (Individual level – Mean)/Standard deviation
> - **Z score** = Observed – Expected / Standard deviation

SKEW DISTRIBUTION

- **Asymmetric Distribution.**
- It is a measure of asymmetry of a probability distribution of random variable.
- One tail of the curve is longer than other.
- **A relative measure of skewness as given by Karl Pearson is:**
  Skewness = [3(Mean – Median)] / SD.

If the skewness so calculated

- **"+ve" → Mean > Median** → Skewness is to the right (Tail is towards right side and there are large number of low scores, small number of high scores)
- **"–ve" → Mean < Median** → Skewness is to the left (Tail is towards left side and there are large number of high scores, small number of low scores)

Distribution	Central Tendency
Normal (Gaussian) distribution	Mean = Median = Mode (coincide)
Right (Positive) skew distribution	Mean > Median > Mode
Left (Negative) skew distribution	Mean < Median < Mode

- Pearson's mode or First Skewness coefficient = (Mean – Mode)/SD
- Pearson's median or Second Skewness coefficient = 3(Mean – Median)/SD
- Quartile skewness = (Q3 – 2Q2 + Q1)/ Q3 – Q1

Negatively skewed distribution (Asymmetric distribution)	Standard normal distribution	Positively skewed distribution (Asymmetric distribution)
Left skewed (or left tailed distribution)	Symmetric distribution	Right skewed (or) Right tailed distribution
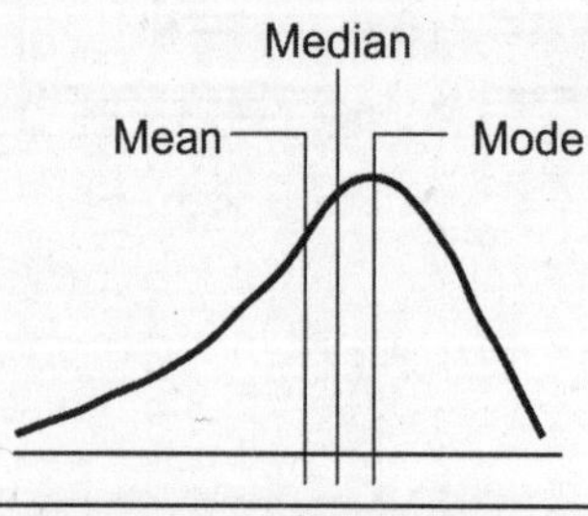	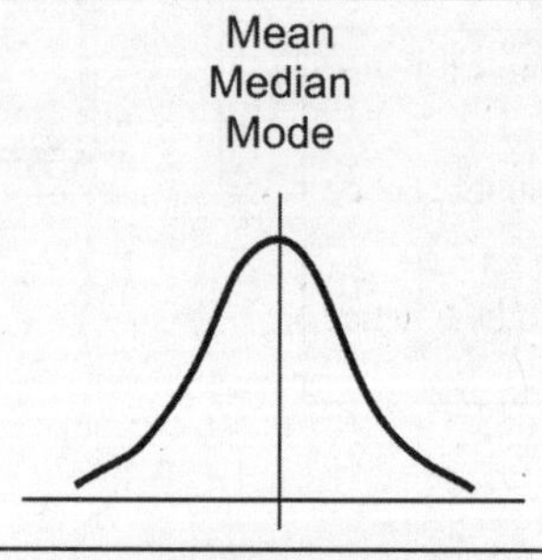	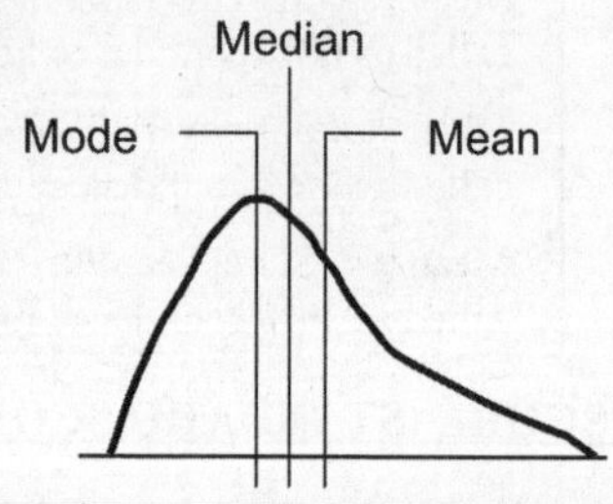
Mean < Median < Mode	**Mean = Median = Mode**	**Mean > Median > Mode**

MULTIPLE CHOICE QUESTIONS

1. **Which is the best distribution to study the daily admission of head injury patients in trauma care centre? [AIIMS 2008]**

   (a) Normal distribution (b) Binomial distribution

   (c) Uniform distribution (d) Poisson distribution

Ans: (d)

POISSON DISTRIBUTION

- It is a 'discrete probability distribution' that expresses 'probability of a number of events occurring in a fixed period of time' (if these events occur with a known average rate and independently of the time since the last event)
- Can also be used for number of events in other specified intervals such as distance, area or volume
- Generally used to model the number of events occurring within a given time interval

2. Standard normal distribution: **[AIPGME 05, AIIMS Nov 99]**

(a) Skewed to the left
(b) Has mean = 1.0
(c) Has standard deviation = 0.0
(d) Has variance = 1.0

Ans: (d)

- SD = σ.
- Variance = σ^2.
- **In Normal distribution, SD = 1, thus Variance = 1.**

3. A normal distribution curve depends on: **[NEET Pattern 2010, AIIMS 2007]**

(a) Mean and sample size
(b) Range and sample size
(c) Mean and standard deviation
(d) Mean and median

Ans: (c)

- Normal/Gaussian curve is 'bilaterally symmetrical, bell-shaped' and based on mean and standard deviation.

4. Fasting blood levels of glucose for a group of diabetics is found to be normally distributed with a mean of 105 mg per 100 ml of blood and a standard deviation of 10 mg per 100 ml of blood. From this data it can be inferred that approximately 95% of diabetics will have their fasting blood glucose levels within the limits of: **[AIIMS 2013]**

(a) 75 and 135 mgs
(b) 85 and 125 mgs
(c) 95 and 115 mgs
(d) 65 and 145 mgs

Ans: (b)

- **Mean = 105 mg/dl, SD = 10 mg/dl.**
- 95% of diabetics will have their fasting blood glucose levels within the limits of Mean ± 2SD within 105 + (2 × 10).
- So it lies between 85 – 125 mg per 100 ml.

5. In a group of 100 children, mean weight of children is 15 kg. Standard deviation is 1.5 kg. Which one of the following is true? **[AIIMS 2017]**

(a) 95% of all children weight between 12 and 18 kg
(b) 95% of all children weight between 13.5 and 16.5 kg

(c) 99% of all children weight between 12 and 18 kg

(d) 99% of all children weight between 13.5 and 16.5 kg

Ans: (a)

- 95% of value are contained in Mean ± 2SD.
- So 95% of all children will have weight between 15 + 2 (1.5), i.e., between 12 and 18 kg.

6. Which of the following statements is incorrect about normal distribution? [AIIMS 2017]

(a) Shows a 'bath tub distribution' (b) Mean = 0

(c) Bilaterally symmetrical bell shaped (d) Variance = 1.0

Ans: (a)

BATH TUB DISTRIBUTION

- It is a distribution where the curve shows trough (depression) in the middle instead of peak.
- **Example:** No. of deaths in various age groups in India (Deaths in India are high in infancy and above the age of > 50 years).

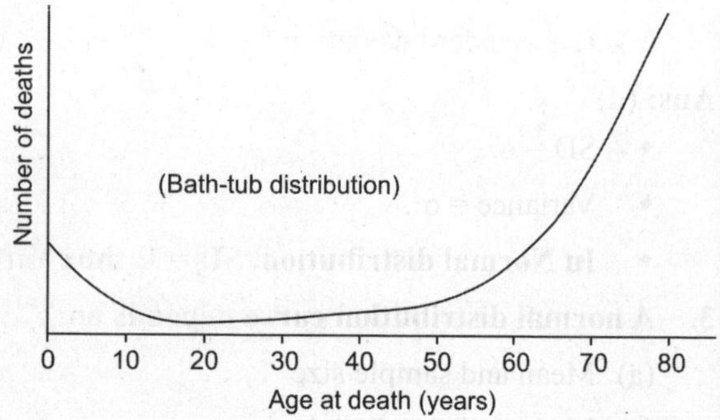

7. For a negatively skewed data mean will be: [AIIMS May 02, 05; NEET Pattern 2013]

(a) Less than median (b) More than median

(c) Equal to median (d) One

Ans: (a)

Distribution	Central Tendency
Normal (Gaussian) distribution	Mean = Median = Mode (coincide)
Right (Positive) skew distribution	Mean > Median > Mode
Left (Negative) skew distribution	Mean < Median < Mode

8. Distribution of random blood glucose measurements from 50 first year medical students was found to have a mean of 3.0 mmol/litre with a standard deviation of 3.0 mmol/litre. Which of the following is a correct statement about the shape of the distribution of random blood glucose in these first year medical students? [AIIMS Nov 2005]

(a) Since both mean and standard deviation are equal, it should be a symmetric distribution

(b) Distribution is likely to be positively skewed

(c) Distribution is likely to be negatively skewed

(d) Nothing can be said conclusively

Ans: (d)

- Mean = 3.0 mmol/litre and Standard deviation = 3.0 mmol/litre.
- Thus, nothing can be said conclusively.

9. **A chest physician observed that the distribution of forced expiratory volume (FEV) in 300 smokers had a median value of 2.5 litres with the first and third quartiles being 1.5 and 4.5 litres respectively. Based on this data how many persons in the sample are expected to have a FEV between 1.5 and 4.5 litres? [AIIMS Nov 05]**

(a) 7.5 (b) 150

(c) 225 (d) 300

Ans: (b)

- Thus Q3 – Q1 will cover 50% values, i.e., 50% of 300 = 150 values.

10. **If the distribution of intra-ocular pressure (IOP) seen in 100 glaucoma patients has an average 30 mm with a SD of 1.0, what is the lower limit of the average IOP that can be expected 95% of times? [AIIMS 2005, AIPGME 2014]**

(a) 28 (b) 26

(c) 32 (d) 29

Ans: (a)

- Mean = 30 mm and SD = 1.0 mm.
- Thus, 95% values are contained in the range of Mean ± 2 SD= 30 ± 2 (1).
- So, 95% values are contained in range 30 – 2 mm and 30 + 2 mm = between 28 and 30 mm.

11. **How much of the sample is included in 1.95 SD? [AIIMS May 1995, AIPGME 1997]**

(a) 99% (b) 95%

(c) 68% (d) 65%

Ans: (b)

Normal (Gaussian) Distribution

- Mean ± 1SD (m ± 1s) limits include 68.27% values
- Mean ± 2SD (m ± 2s) limits include 95.45% values
- Mean ± 1.96SD (m ± 1.96 m) limits include 95% values
- Mean ± 3SD (m ± 3s) limits include 99.73% values
- Mean ± 2.58SD (m ± 2.58 m) limits include 99% values

12. **If the systolic blood pressure in a population has a mean of 130 mm Hg and a median of 140 mm Hg, the distribution is said to be: [NEET Pattern 2014]**

(a) Symmetrical

(b) Positively skewed

(c) Negatively skewed

(d) Either positively or negatively skewed depending on the Standard deviation

Ans: (c)

- Mean systolic blood pressure in population (130 mm Hg) is less than median (140 mm Hg).
- Thus distribution is Left (Negative) skew distribution.

13. For given set of values, Mean = 20, Median = 24 and Mode = 26. Given distribution is: **[NEET Pattern 2016]**

(a) Symmetric (b) Right-skewed

(c) Left-skewed (d) Can be either symmetric or skewed

Ans: (c)

- Mean = 20, Median = 24 and Mode = 26.
- Thus Mean < Median < Mode, making it a Left (Negative) skew distribution.

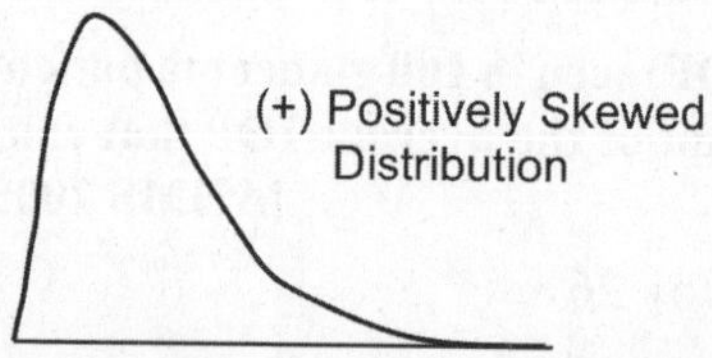

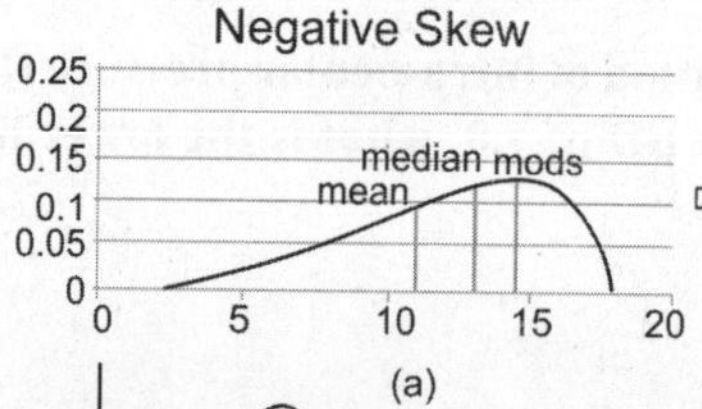

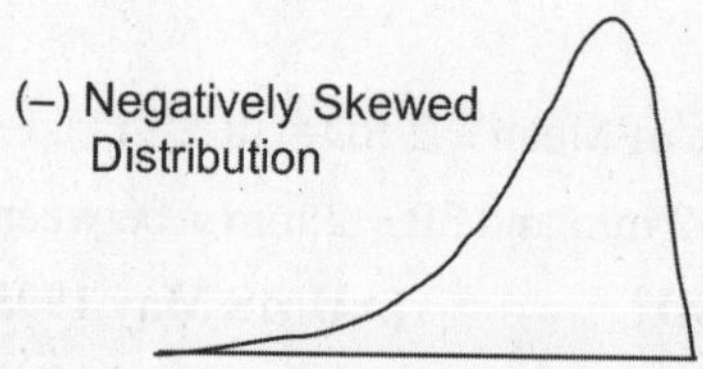

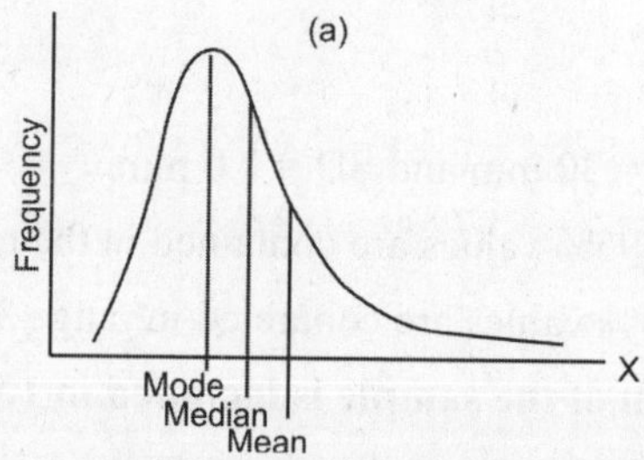

14. A population study showed a mean glucose of 86 mg/dL. In a sample of 100 showing normal curve distribution, what percentage of people have glucose above 86 mg/dL? **[AIIMS 2015, AIPGME 2012]**

(a) 34% (b) 50%

(c) 90% (d) 68%

Ans: (b)

- Mean glucose = 86 mg/dL.
- Thus, 50% of people will have glucose above 86 mg/dL and 50% of people will have glucose below 86 mg/dL.

15. Systolic blood pressure of a group of normal people between the age of 25-27 years was taken and a mean of 120 mm Hg was found. What will be the expected number of individuals among a group of 100 people taken for study whose systolic blood pressure will be below 120 mm Hg? **[AIIMS 2016]**

(a) 25% (b) 50%

(c) 75% (d) 100%

Ans: (b)

- Mean Systolic blood pressure (SBP) = 120 mm Hg.

- So 50% of values (50% of n), i.e., 50 individuals will have their SBP below 120 mm Hg, and a similar no. will have value of SBP above 120 mm Hg.

16. PEFR of group of 11-year-old girls follow Normal distribution with Mean 300 Litre/min and Standard deviation 20 Litre/min: [AIPGME 05]

(a) About 95% of the girls have PEFR between 260 and 340 L/min

(b) Girls have healthy lungs

(c) About 5% of girls have PEFR below 260 L/min

(d) All the PEFR must be less than 340 L/min

Ans: (a)

- 95% of girls will have PEFR in range Mean ± 2SD = 300 + 2(20) l/min, i.e., between (300 –- 40) and (300 + 40) l/min.
- 260 – 340 l/min range covers 95% of girls.

17. In normal curve: [Karnataka 2006]

(a) Mean = 2 standard deviation (b) Mean = Median

(c) Mean = Variance (d) Mean = 1 standard deviation

Ans: (b)

- **MEAN = MEDIAN = MODE.**

18. Z score is for which type of distribution? [NEET Pattern 2010]

(a) Normal (b) Binomial

(c) t test (d) Chi-square test

Ans: (a)

- **Z score is for NORMAL distribution.**

19. Pearson's Skewness Coefficient is given by: [NEET Pattern 2011]

(a) Mean–Mode/SD (b) Mode–Mean/SD

(c) SD/Mean–Median (d) SD/Median–Mean

Ans: (a)

- Pearson's mode or First Skewness coefficient = (Mean – Mode)/SD.
- Pearson's median or Second Skewness coefficient = 3(Mean – Median)/SD.

20. Mean value of marks in a distribution of 100 students in a class is 105 and Standard deviation is 10. How many students will have their marks in the range 85-125? [NEET Pattern 2012]

(a) 50% (b) 68%

(c) 95% (d) 99.7%

Ans: (c)

- Mean ± 2SD = 105 ± 20 = 85-125 marks covers 95% students.
- Mean ± 3SD = 105 ± 30 = 75-135 marks covers 99% students.

21. When the variables follow standard distribution: **[DNB Pattern 2008]**

(a) Mean = median
(b) Median = variance
(c) Mean = 2 median
(d) Standard deviation = 2 variance

Ans: (a)

- Refer above explanation.

22. In a left skewed curve, true statement is: **[DNB June 2011]**

(a) Mean = Median
(b) Mean < Mode
(c) Mean > Mode
(d) Mean = Mode

Ans: (b)

- Refer above explanation.

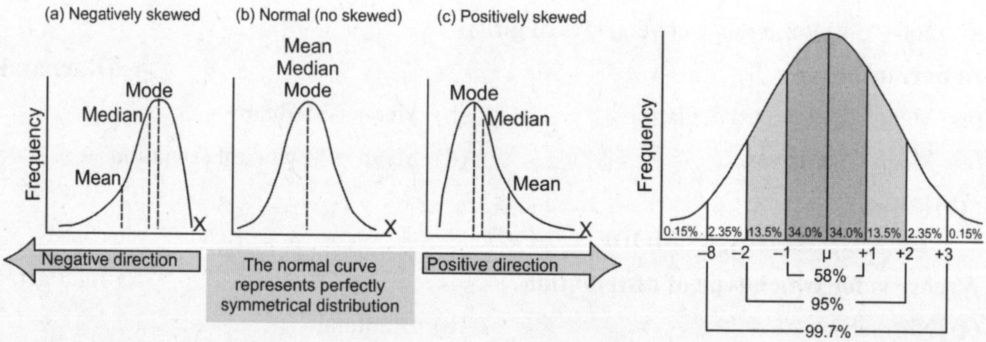

23. Mean hemoglobin of a group of pregnant females is found to be 10.6 gm/dL with Standard deviation of 2 gm/dL. 5% pregnant females in this group will have their hemoglobin level below: **[AIIMS May 2014]**

(a) 8.6 gm/dL
(b) 7.31 gm/dL
(c) 6.6 gm/dL
(d) 5.0 gm/dL

Ans: (b)

- Mean + 1SD covers 68% values. So, Hb of 68% pregnant females will lie between 10.6 + 2 gm/dL or between 8.6-12.6 gm/dL. So, 16% will lie below 8.6 gm/dL.
- Mean + 2SD covers 95% values. So, Hb of 95% pregnant females will lie between 10.6 + 2(2) gm/dL or between 6.6-14.6 gm/dL. So, 2.5% will lie below 6.6 gm/dL.
- **<u>So, 5% pregnant females will have Hb below 7.31 gm/dL.</u>**

24. In a standard normal curve, area between one standard deviation on either side will be: **[DNB 2000]**

(a) 68%
(b) 85%
(c) 99.7%
(d) None of the above

Ans: (a)

- One standard deviation= 68%.

25. In a standard normal curve, mean + 2 standard deviations covers: **[DNB Pattern 2000]**

(a) 60% (b) 65%

(c) 95% (d) 99%

Ans: (c)

- **2 Standard deviation is equal to 95%.**

26. When the variables follow standard distribution: **[DNB Pattern 2008]**

(a) Mean = median (b) Median = variance

(c) Mean = 2 median (d) Standard deviation = 2 variance

Ans: (a)

- Mean = median = mode.

27. About standard Normal curve all is true *except*: **[NEET Pattern 2012]**

(a) Area = 1 (b) SD = 1

(c) Bell shaped (d) Mean and Median = 1

Ans: (d)

- In a normal distribution, Mean = 0 and SD = 1.
- **Mean = median = mode = ZERO.**

28. Area covered between two standard deviation in a normal distribution curve: **[NEET Pattern 2014]**

(a) 68% (b) 95.4%

(c) 99.6% (d) 100%

Ans: (b)

- 2 standard deviation is equal to 95% confidence limits.

29. Normal distribution curve is determined by: **[Rajasthan 2002, MH 2006]**

(a) Standard deviation and mean (b) Standard deviation and mode

(c) Mode and median (d) Standard deviation and median

Ans: (a)

- Refer above explanation.

NEW PATTERN MCQs

1. Which of the following is true about Standard deviation?

(a) Skewed to left (b) Mean-1

(c) Variance-1 (d) SD-zero

Ans: (d)

- Bell shaped smooth curve.

- Symmetrical.
- Both ends doesn't touch baseline.
- Variance – 1, SD = 1 → **Variance = (S.D)²**.

2. **Distribution of Random blood glucose measurements from 30 individuals was found to be 120+/- 10 mg/dL. Which of the following is a correct statement about shape of distribution of RBS in these first year medical students?**
   (a) Since both mean and SD are equal should be symmetric distribution
   (b) Distribution is likely to be +ve skewed
   (c) Distribution is likely to be -ve skewed
   (d) Nothing can be concluded

Ans: (d)

- Median and mode not provided. So it can't be concluded regarding skewness.

EXTRA EDGE FOR INSTITUTE BASED EXAMS

POISSON DISTRIBUTION:

- Used to describe occurrence of rare events in a large population.
- Probability of occurrence of event is indefinitely small and 'n' the number of observations is very large.
- Probability distribution for count of events (earthquake) occurring randomly in a given period of time.
- Here variance = mean.

BINOMIAL DISTRIBUTION:

- Distribution of outcome from a series of data characterized by 2 mutually exclusive [Dichotomous] categories.
- Population under observation can be divided into 2 groups, one with a certain characteristic and other without it.
- If the probability of a person being male is "p" and not being male is "q" [1-p] in a group of "n" persons.
- Then mean number of frequencies of males = np, SD = npq.
- If p = q, the curve will be symmetrical or else skewed.

UNIFORM DISTRIBUTION/RECTANGULAR DISTRIBUTION:

- A distribution in which all events occur with equal frequency. It does not occur very often.
- Uniform distribution divided into equally spaced intervals, will have equal number of members in each interval.

LOG NORMAL DISTRIBUTION:

- It is a skewed distribution when graphed using a arithmetic scale but a normal distribution when graphed using a logarithmic scale.

CHAPTER

12

CONFIDENCE INTERVALS

CONFIDENCE INTERVAL (CI):

- It is the estimated range of values in a sample, within which a population parameter value is expected to lie with certain confidence levels (95% or 99%).
- It is expressed as a range that is likely to contain the population parameter when so obtained for repeated samples.
- **100 percent Confidence interval can never be achieved.**
- A narrow CI is always preferable: as it be more nearer to the population parameter.
- **Formula for 95% CI = Mean ± 1.96 SE.**
- **Larger the sample size, narrower is CI; Smaller the Standard deviation, narrower is CI.**
- Confidence level = $1 - \alpha$.

Confidence limits	Normal deviate (N.D.)	Significance
μ is outside the 95% confidence limits	N.D. > 2	P < 0.05, Significant at 5% level
μ is just within 95% confidence limits	N.D. = 2	P = 0.05, Just significant at 5% level
μ is within the 95% confidence limits	N.D. < 2	P > 0.05, Significant at 5% level

CONFIDENCE LIMITS:

- They are upper and lower boundaries of range is expressed as confidence interval.

MULTIPLE CHOICE QUESTIONS

1. **Confidence limit in 2 S.D. is:** **[NEET Pattern 2014]**

   (a) 66% (b) 95%

   (c) 98% (d) 90%

Ans: (b)

- Refer to above explanation.

2. **Confidence limit is calculated by using:** **[NEET Pattern 2016]**

   (a) Mean and standard error (b) Mean and standard deviation

   (c) Median and standard deviation (d) Median

Ans: (b)

- It can be achieved by mean and standard deviation.
- Larger the sample size, Smaller the Standard deviation = narrower is CI.

3. If a 95% Confidence interval for prevalence of cancer in smokers > 65 years is 56 to 76% Chance that prevalence could be less than 56% is:

(a) Nil (b) 4%

(c) 2.5% (d) 5%

Ans: (c)

- If 95% is within 56 to 76%.
- Then rest of 5% will be on either side of distribution.
- So, 2.5% probability will have < 56%.

4. What is the probability that confounding factor falls to right of 95%?

(a) 1 in 10 (b) 1 in 20

(c) 1 in 30 (d) 1 in 40

Ans: (d)

- 2.5% lies to right of 95% SD.
- Therefore it is 1 in 40.

SUMMARY

Confidence Level	Multiplier
68%	1
90%	1.6
95%	1.96
99%	2.58

CHAPTER 13

TESTS OF SIGNIFICANCE

They are of 2 types of test of significance

- Parametric tests.
- Non-Parametric tests.

Condition	Use Parametric Tests	Use Non-Parametric Tests
Type of sampling used	Random sampling	Non-probability sampling
Sample size	Is calculated/Large	Is not calculated/small
Distribution of data	Normal distribution (Mean = Median = Mode)	Skewed distribution (not normal) (Mean > Median > Mode (or) Mean < Median < Mode)
Type of data	Quantitative data	Qualitative data Quantitative data (skewed)
Type of variables	Discrete or Continuous	Categorial Discrete or Continuous (skewed)
Scale of measurements	Interval or Ratio scale data, if normally distributed	Nominal or Ordinal scale data, Interval or ratio scale, if not normally distributed
Relationship between Standard Deviation (S.D.) and Mean	S.D. < ½ of Mean	S.D. > ½ of Mean
Measures of Central Tendency used for comparison	Mean and S.D.	Median and Inter-quartile Range (IQR = Q3-Q1)
Example	**Parametric Tests:** Z test Unpaired t test Paired t test One way ANOVA Two way ANOVA	**Non-Parametric Tests:** Chi square test Wilcoxon rank sum test Wilcoxon signed rank test Kruskal Wallis test Friedman's test

PARAMETRIC TESTS

Student 't' test [Devised by WS Cassette]

- Based on ***"t" distribution*** [a continuous, symmetrical, bell shaped, unimodal distribution of infinite range similar in shape to normal distribution, but more spread out].
- It shows greater variation due to chance than normal distribution and is used to analyse small samples.
- ***As sample size increases, t distribution closely resembles normal distribution.***

- ***As sample size decreases, t value decreases and probability that difference is ascribed to chance increases and there is increased probability for accepting null hypothesis.***

> - **Student t test for single small sample**–Compares a single sample mean with population mean.
> - Student t test for independent samples **(Unpaired t test)**–Compares mean of 2 small samples [< 30 size].
> - Student t test for paired samples **(Paired t test)**–Compares mean of 2 small paired samples [< 30 size].
> - **Z - test: Is a variant of student's t-test which is used when sample size is > 30.**

- Each individual gives a pair of observation, may be before and after play of a factor, e.g., BP before and after drug intake.
- Used to evaluate the null hypothesis for continuous (quantitative) variables (height, BP).
- ***Applied to find significance of difference between two means, when sample size is < 30.***
- For each degree of freedom, a row of increasing t value corresponds to a row of decreasing probability for accepting the null hypothesis.

ANOVA TEST (F-test/F-ratio):

- Comparing means (+ SD) in more than two different group of individuals.

NON PARAMETRIC TESTS

CHI- SQUARE TEST (X^2)–DEVELOPED BY KARL PEARSON:

- It is a type of non-parametric test *used for qualitative data* in form of proportions.
- It is an alternate method of testing the significance of difference between two proportions.
- **Used as a test of goodness of fit:** It determines whether the observed frequency distribution differs from theoretical distribution [i.e., normal, binomial or Poisson] by chance or sample is drawn from a different population.
- Used to find whether any association between 2 events/discrete attributes is real/chance.
- *Advantage*: Can also be used when more than two groups are to be compared.
- ***Yates correction:*** *Applied when expected frequency in any cell of 2 × 2 contingency table is < 5.*

WILCOXON TESTS:

- **Wilcoxon sign rank test:** Comparing percentage, proportions and fractions in **matched paired data.**
- **Wilcoxon rank sum test:** Comparing percentage, proportions and fractions in two **unpaired samples.**

MANN-WHITNEY TEST:

- It is same as Wilcoxon rank sum test.
- It is used for assessing whether two set of observations come from same distribution (if 2 independent 'nonpaired' samples come from the same population).

- It is analogous to parametric 2 sample 't-test' on data after ranking over combined samples.

Mc NEMAR'S TEST:

- It is a non-parametric test used on nominal data.
- Applied to 2 × 2 contingency table with dichotomous trait, with matched pairs of subjects.

DEGREES OF FREEDOM

- **Degree of freedom:** No. of observations in a data set that can freely vary once parameters have been estimated.
  - **Chi-square test,** contingency table, dof = (c – 1) (r – 1) (where c = no. of columns and r = no. of rows)
  - **Student's t-test** (one-sample data/paired test), dof v = n – 1 (where n = no. of units in the sample)
  - **Student's t-test** (two-sample data/unpaired test), dof v = (n1 + n2) – 1 (where n1 and n 2 = no. of units in the two samples).

SUMMARY

Condition	Dependent Variable	Parametric Test	Non-Parametric Test
Comparisons between 2 independent groups	Continuous	Z test (n > 30) unpaired t test (n ≤ 30)	Wilcoxon rank sum test (Mann Whitney 'U' test
	Discrete	Z test (n > 30) unpaired test (n ≤ 30)	Chi square (χ^2 test)
Comparisons between 2 paired groups	Continuous	Paired t test	Wilcoxon signed rank test
	Discrete	Paired t test	McNemar's (χ^2 test)
Comparisons among > 2 independent groups	Continuous	One way ANOVA	Kruskal Wallis 'H' test
	Discrete	One way ANOVA	Chi square (χ^2 test)
Comparisons among > 2 paired groups	Continuous	Repeated measures ANOVA	Friedman's test
	Discrete	Repeated measures ANOVA	Mc Nemar's (χ^2 test)
Correlation between 2 quantitative variables	Continuous/Discrete	Pearson's correlation coefficient (r)	Spearman's correlation coefficient (ρ)

MULTIPLE CHOICE QUESTIONS

1. In a study, variation in cholesterol was seen before and after giving a drug. Test which would give its significance is: [AIPGME 01, 02, 07]

(a) Unpaired t-test (b) Fischer test

(c) Paired t-test (d) Chi-square test

Ans: (c)

- This test is done before and after drug administration.
- Since cholesterol is a quantitative variable.
- Paired t test should be used in this case.

	Normal distribution	Non-normal distribution
Describe one group	Mean, SD	Median, interquartile range
Compare one group to a hypothetical value	One-sample *t* test	Wilcoxon test
Compare two unpaired groups	Unpaired *t* test	Mann-Whitney test
Compare two paired groups	Paired *t* test	Wilcoxon test
Compare three or more unmatched groups	One-way ANOVA	Kruskal-Wallis test
Compare three or more matched groups	Repeated-measures ANOVA	Friedman test
Quantify association between two variables	Pearson correlation	Spearman correlation
Predict value from another measured variable	Simple linear regression or Non-linear regression	Non-parametric regression
Predict value from several measured or binomial variables	Multiple linear regression or Multiple non-linear regression	

2. A cardiologist waewnts to study the effect of an atrovastatin drug. He notes down initial cholesterol levels of 50 patients and then administers the drug on them. After a month's treatment, he measures cholesterol level again. Which of the following is most appropriate to test statistical significance of change in blood cholesterol? [AIPGME 02]

(a) Paired t-test
(b) Unpaired or independent t-test
(c) Analysis of variance
(d) Chi-square test

Ans: (a)

- Before and after event–Paired t test.

3. A chi-square test would be most appropriate for testing which one of following Hypotheses? [AIPGME 2000]

(a) Mean Exam score of Delhi students is greater than that of Mumbai students
(b) Smaller proportion of people who were immunized against chickenpox subsequently develop zoster than those who were not immunized
(c) Mean blood pressure of black and white male-hypertensive patients taking ACE inhibitors is same as that of black and white female-hypertensive patients taking ACE inhibitors and that of black and white males and females taking diuretics and placebos
(d) Mean cost of treating a patient with coronary artery disease with angioplasty is greater than mean cost of providing medical treatment

Ans: (b)

EVALUATION OF TESTS OF SIGNIFICANCE
• **Option A:** Mean score of 2 groups are compared–most appropriate test of significance is Students Unpaired t-test
• **Option B:** Two proportions are compared–most appropriate test of significance is Chi-square test
• **Option C:** Mean score of 3-4 groups are compared–most appropriate test of significance is ANOVA test
• **Option D:** Mean score of 2 groups are compared–most appropriate test of significance is Unpaired Students t-test

4. **In a particular trial, the association of lung cancer with smoking is found to be 40% in one sample and 60% in another. What is the best test to compare results? [AIIMS 2001]**

(a) Chi-Square Test (b) Fischer Test

(c) Paired t Test (d) ANOVA Test

Ans: (a)

- Association of lung cancer with smoking is found to be 40% in one sample and 60% in another.
- Since two proportions are to be compared best test will be Chi-square test.

APPLICATIONS OF CHI-SQUARE TEST

- Test of proportions
- Test of association
- Test of goodness of fit

5. **Height of 20 boys aged 10 years was 140 + 13 cm and 20 girls of same age was 135 cm + 7 cm to test statistical significance of difference in height, test applicable is: [AIIMS 05]**

(a) Chi-square test (b) Z test

(c) t test (d) F test

Ans: (c)

- **Paired Student's t-test:** Comparing means (+ SD) in paired data (in same group of individuals before and after an intervention).
- **Unpaired Student's t-test:** Comparing means (+ SD) in two different group of individuals.
- **Z-test:** It is a variant of student's t-test which is used when sample size is **> 30.**

6. **Mean B.P. of a group of persons was determined and after an interventional trial, mean BP was estimated again. The best test to be applied to determine significance of intervention is: [AIIMS 2007]**

(a) Chi-square (b) Paired 't' test

(c) Correlation coefficient (d) t-test

Ans: (b)

- **Refer to above question for explanation.**

7. **An investigator wants to study the association between maternal intake of iron supplements (Yes or No) and incidence of low birth weight (< 2500 or > 2500 gms). He collects relevant data from 100 pregnant women as to the status of usage of iron supplements and the status of low birth weight in their newborns. The appropriate statistical test of hypothesis advised in this situation is: [AIIMS 2013]**

(a) Paired t-test (b) Unpaired or independent t-test

(c) Analysis of variance (d) Chi-square test

Ans: (d)

- Chi-square test as a test of association between 2 events in binomial or multinomial samples is its most important application.

8. While applying chi-square test to a contingency table of 4 rows and 4 columns, the degrees of freedom would be: [AIPGME 2005]

(a) 1 (b) 4

(c) 9 (d) 8

Ans: (c)

- For chi-square test, Degree of freedom = (c – 1) (r – 1)
- Therefore (4 – 1) (4 – 1) = 9

9. In a 3 × 4 contingency tables, the number of degrees of freedom equals to: [AIIMS 2004]

(a) 1 (b) 5

(c) 6 (d) 12

Ans: (c)

- **Chi-square test,** contingency table, dof = (c – 1) (r – 1) (where c = no. of columns and r = no. of rows).
- **Student's t-test** (1 sample data/paired test), dof v = n – 1 (where n = no. of units in sample).
- **Student's t-test** (two-sample data/unpaired test), dof v = (n1 + n2) – 1 (where n1 and n 2 = no. of units in two samples).

10. A cardiologist wants to study effect of an anti-hypertensive drug. He notes down initial systolic blood pressure (mm Hg) of 50 patients and then administers drug on them. After a week's treatment, he measures the following is the most appropriate statistical test of significance to test the statistical significance of the change in blood pressure: [AIIMS 1997, AIIMS 1995, AIIMS 2004]

(a) Paired t-test (b) Unpaired or independent t-test

(c) Analysis of variance (d) Chi-square test

Ans: (a)

- **Before and after administration of drug.** So Paired t test should be used.

11. A study was undertaken to assess effect of a drug in lowering serum cholesterol levels. 15 obese women and 10 non-obese women formed the 2 limbs of the study. Which test would be useful to correlate the results obtained? [AIIMS 2001]

(a) ANOVA test (b) Student's t-test

(c) Chi-square test (d) Fischer test

Ans: (b)

Student's *t*-Test

- **Paired Student's t-test:** Comparing means/SD in paired data (in same group of individuals before and after an intervention)
- **Unpaired Student's t-test:** Comparing means/SD in 2 different group of individuals
- **Z-test:** Is a variant of student's t-test which is used when sample size is > 30

12. In a particular trial, the association of lung cancer with smoking is found to be 40% in one sample and 60% in another. What is the best test to compare the results? [AIPGME 03]

(a) Chi-square test (b) Fischer test

(c) Paired t test (d) ANOVA test

Ans: (a)

- Since two proportions are to be compared best test will be Chi-square test.

Applications of Chi-Square Test

- Test of proportions
- Test of association
- Test of goodness of fit

13. Not required for Chi-square test is: [AIIMS Dec 1997]

(a) Mean and SD of the groups (b) Each expected cell frequency > 5

(c) Large sample (d) Contingency Table

Ans: (a)

Applications of Chi-Square Test

- Test of proportions
- Test of association
- Test of goodness of fit

14. Appropriate statistical method to compare two means is: [NEET Pattern 2000, 2013]

(a) Chi-square test (b) Student's t-test

(c) Odds Ratio (d) Correlation Coefficient

Ans: (b)

- **Paired Student's t-test:** Comparing means/SD in paired data (in same group of individuals before and after an intervention).
- **Unpaired Student's t-test:** Comparing means/SD in 2 different group of individuals.

15. Appropriate statistical method to compare two proportions is: [AIPGME 2005]

(a) Chi-square test (b) Student's t-test

(c) Odds Ratio (d) Correlation Coefficient

Ans: (a)

Applications of Chi-Square Test

- Test of proportions
- Test of association
- Test of goodness of fit

16. In a given data, degree of freedom will be: **[AIIMS 2006]**

Duration of developing AIDS	Blood group			
	A	B	AB	O
0 – 5 years	20	30	48	7
5 – 10 years	110	12	37	12
10 – 15 years	12	9	8	3

(a) 12 (b) 6

(c) 9 (d) 20

Ans: (b)

- There are 3 rows and 4 columns.
- Thus, dof = (c – 1) (r – 1) = (4 – 1) (3 – 1) = 6.

17. For testing statistical significance of difference in heights of school children: **[AIIMS 2003]**

(a) Student's 't' test (b) Chi-squared test

(c) Paired 't' test (d) One way analysis of variance

Ans: (d)

ANOVA test (F-test/F-ratio):

- Comparing means (+ SD) in more than two different group of individuals.

18. Not true about Chi-square test is: **[AIPGME 2003, AIIMS 2009]**

(a) Tests the significance of difference between two proportions

(b) Tells about presence or absence of an association between two variables

(c) Directly measures the strength of association

(d) Can be used when more than two groups are to be compared

Ans: (c)

Applications of Chi-Square Test
• Test of proportions
• Test of association
• Test of goodness of fit

19. Mean bone density amongst 2 group of 50 people each is compared, which is the best test? **[AIIMS May 2008]**

(a) Chi-square (b) Student t test

(c) Mc Nemar chi-square test (d) Fischer test

Ans: (b)

- Unpaired Student's t-test is most appropriate to test significance of association.

20. An investigator finds out that 5 independent factors influence the occurrence of a disease. Comparison of multiple factors that are responsible for the disease can be assessed by: [AIIMS May 2011]

(a) ANOVA
(b) Multiple linear regression
(c) Chi-square test
(d) Multiple logistic regression

Ans: (d)

- **Multiple linear regression:**

  It is used if the target variables are dimensional having multiple possible values.

- **Multiple logistic regression:**

  It is used if the target variables are binary having only two possible values.

21. Mean blood alcohol levels are measured in patients before and after using an interventional drug. Statistical test of significance to be applied is: [AIPGME 2012]

(a) Chi-square test
(b) ANOVA
(c) Paired students t-test
(d) Unpaired students t-test

Ans: (c)

- **Before and after intervention**–Paired t test is therefore used.

22. Tests of Significance include all *except*: [JIPMER 2014]

(a) t
(b) Z
(c) SD
(d) Chi-square

Ans: (c)

- Mean and SD are measure of central tendency.

23. Chi-square test is used to measure the degree of: [Bihar 2004]

(a) Significance of difference between two proportions
(b) Association between two variables
(c) Correlation between two variables
(d) Agreement between two observations

Ans: (a)

Applications of Chi-Square Test
• Test of proportions
• Test of association
• Test of goodness of fit

24. True regarding Chi-square test is: [UP 2000]

(a) Measure Significance of difference between two proportion
(b) Null hypothesis is equal

(c) Does not test the significance

(d) Tests correlation and regression

Ans: (a)

Applications of Chi-Square Test

- Test of proportions
- Test of association
- Test of goodness of fit

25. In a table of 2 × 2, the degree of freedom is: **[UP 2001]**

(a) 4
(b) 1
(c) 8
(d) 7

Ans: (b)

- Degree of freedom = (c – 1) (r – 1) = 1

26. Degree of freedom for a contingency table with 3 rows and 6 columns is: **[NEET Pattern 2012]**

(a) 2
(b) 3
(c) 10
(d) 18

Ans: (c)

- Degree of freedom = (c – 1) (r – 1) = 10

27. Test of association between two qualitative variables is done by: **[AP 2004]**

(a) Chi-square test
(b) Correlation
(c) Regression
(d) None

Ans: (a)

Applications of Chi-Square Test

- Test of proportions
- Test of association
- Test of goodness of fit

28. About chi-square test, true is: **[MP 2000]**

(a) < 0.001 is statistically significant

(b) Less no. of samples are associated with less error

(c) Categories of data used in test need not be mutually exclusive and discrete

(d) Tests correlation and regression

Ans: (a)

- **p Value < 0.05 is statistically significant.**

29. A study who planned to find out the effect of iron supplementation during pregnancy on birth weight of newborn children. Two groups, one with iron supplementation and other without

iron supplementation during pregnancy were compared. Birth weight of newborns were recorded as means + SD. Which significant test (statistical) is appropriate for comparison of birth weights between the two groups? **[MP 2008]**

(a) Unpaired 't' test (b) Paired 't' test

(c) McNemar's chi-square test (d) Chi-square test

Ans: (a)

- Two groups are compared by mean values–Unpaired t test.

30. A study measures patient's serum cholesterol before and after new lipid-lowering therapy has been given. What type of significance test should be used to analyze data? **[MH 2007]**

(a) Paired t-test (b) Student's t-test

(c) Chi-squared test (d) Pearson's test

Ans: (a)

- Before and after intervention–Paired test.

31. What would be degree of freedom if 5 row and 6 columns? **[RJ 2009]**

(a) 30 (b) 20

(c) 40 (d) 25

Ans: (b)

- Degree of freedom = (c – 1) (r – 1) = 20

32. Chi-square test 5 rows/4 columns, degree of freedom is: **[NEET Pattern 2012, 2013]**

(a) 9 (b) 12

(c) 16 (d) 20

Ans: (b)

- Degree of freedom = (c – 1) (r – 1) = 12

33. ANOVA is used: **[NEET Pattern 2012]**

(a) To compare means in 2 groups

(b) To compare means in 3 or more groups

(c) To compare means in 1 group before and after intervention

(d) To find correlation

Ans: (b)

ANOVA test (F-test/F-ratio):

- Comparing means (+ SD) in more than two different group of individuals

34. Not required for chi-square test: **[NEET Pattern 2012]**

(a) Null hypothesis (b) Degrees of freedom

(c) Means in different groups (d) Proportions in different groups

Ans: (c)

Applications of Chi-Square Test
• Test of proportions
• Test of association
• Test of goodness of fit

35. Chi-square test is for: **[DNB Pattern 2010]**

(a) Standard error of mean
(b) Standard error of proportion
(c) Standard error of difference between 2 means
(d) Standard error of difference between proportions

Ans: (d)

Applications of Chi-Square Test
• Test of proportions
• Test of association
• Test of goodness of fit

36. Degree of freedom for 2 × 2 contingency table is: **[DNB Pattern 2011]**

(a) 1 (b) Zero
(c) 2 (d) 4

Ans: (a)

- Degree of freedom = (c – 1) (r – 1) = 1

37. Degree of freedom of Chi-square test in contingency table of 2 by 3 is: **[DNB Pattern 2009]**

(a) 1.0 (b) Zero
(c) 2 (d) 4

Ans: (c)

- Degree of freedom = (c – 1) (r – 1) = 2

ATTRIBUTE	TYPE	TEST
Probability distribution	Normal	'Z' test
	Near normal	't' test, Chi-square test, Fishers test
Number of samples to be compared	Two	't' test, Chi-square test, 'Z' test
	More than two	ANOVA
Sample size	Small (≤ 30)	't' test, ANOVA, Non parametric test
	Large (>30)	Z test, Chi-square test
Type of data	Qualitative	Chi-square test, Kruskal-Wallis test
	Quantitative	't' test, ANOVA, Z test

MISCELLANEOUS
• **Kolmogrov smirnov test:** Non-parametric test used to check normality of data. • **Dixon Q test:** Identify and reject outliers. • **Cohen's kappa test:** It is a measure for degree or level of agreement between observers based on qualitative categorical data.

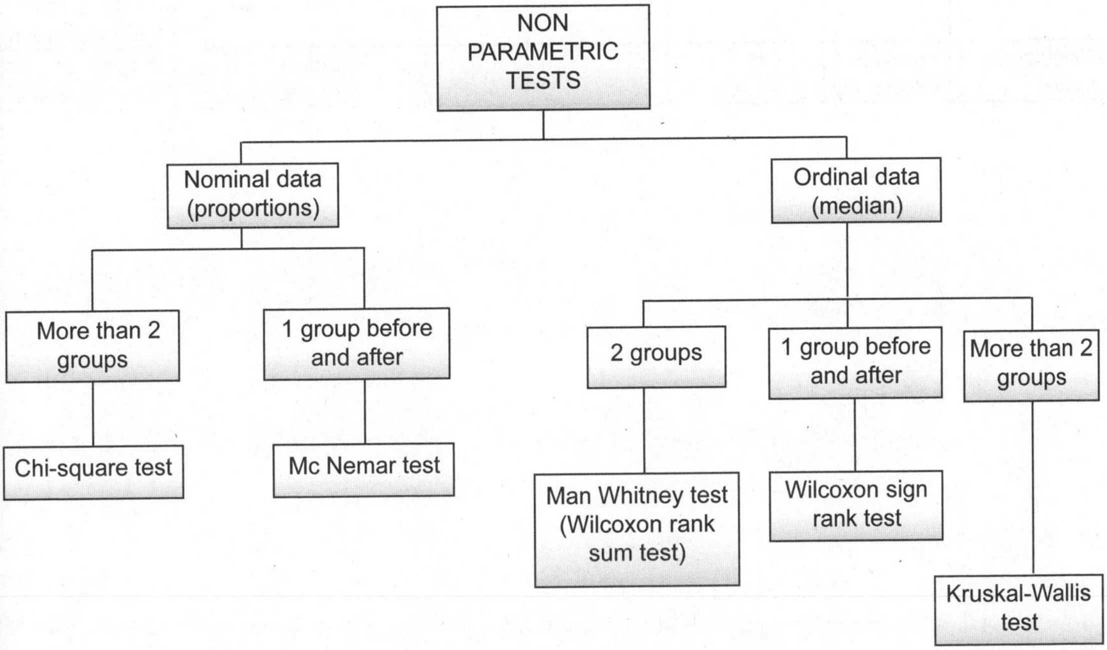

CHOICE OF STATISTICAL TEST FOR INDEPENDENT OBSERVATIONS

	OUTCOME VARIABLE					
Input variable	**Nominal**	**Categorical (>2 categories)**	**Ordinal**	**Quantitative discrete**	**Quantitative non-normal**	**Quantitative normal**
Nominal (e.g. treatment vs placebo)	χ^2 or Fisher's	χ^2	χ^2 trend or Mann-Whitney	Mann-Whitney	Mann-Whitney or log rank	Student's t test
Categorical (>2 categories)	χ^2	χ^2	Kruskal-Wallis	Kruskal-Wallis	Kruskal-Wallis	Analysis of variance
Ordinal (ordered categories)	χ^2 trend or Mann-Whitney		Spearman rank correlation	Spearman rank correlation	Spearman rank correlation	Spearman rank correlation or linear regression

Contd.

Contd.

Quantitative discrete	Logistic regression			Spearman rank correlation	Spearman rank correlation	Spearman rank correlation or linear regression
Quantitative non-normal	Logistic regression				Plot data and Pearson or Spearman rank correlation	Plot data and Pearson or Spearman rank correlation and linear regression
Quantitative normal	Logistic regression				Linear regression	Pearson or linear regression

CHAPTER

14

CORRELATION AND REGRESSION

- It indicates the relation between **2 Quantitative** variables–Discrete and Continuous.
- Change in the value of Independent variable, results in a change in the value of dependent variable.
- **Correlation is represented by: 'Scatter diagram'.**
- **Correlation Coefficient (r)** indicates **degree or strength** of relationship between **2 Quantitative variables.**
- **Direction: Sign** (+ / -) indicates direction of relationship.
- **Strength: 'r' value** indicates the strength of relationship. It ranges between -1 and +1.

DIRECTION OF LINEAR CORRELATION
Positive Correlation (Direct Relation): • As value of Independent variable increases, value of dependent variable also increases • Independent Variable α Dependent Variable • **Examples:** Age and Weight, BMI and SBP, Fat intake and Serum LDL **r = + 1 is a Perfect Positive Correlation**
Negative Correlation (Inverse Relation): • As the value of the Independent variable increases, the value of dependent variable decreases • Independent Variable α (1/Dependent Variable) • **Examples: Insulin vs Plasma Glucose HDL and LDL, Age vs Vital capacity of Lungs** **r = – 1 is a Perfect Negative Correlation**
Zero Correlation or No Correlation: • When 2 variables are independent, i.e., change in one variable has no effect on other • **Examples:** Hb % and Serum LDL, Height and Serum IgG titre

Positive Correlation

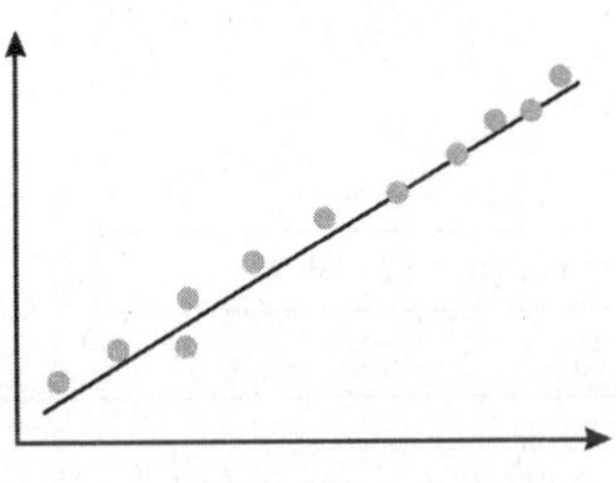

Zero Correlation

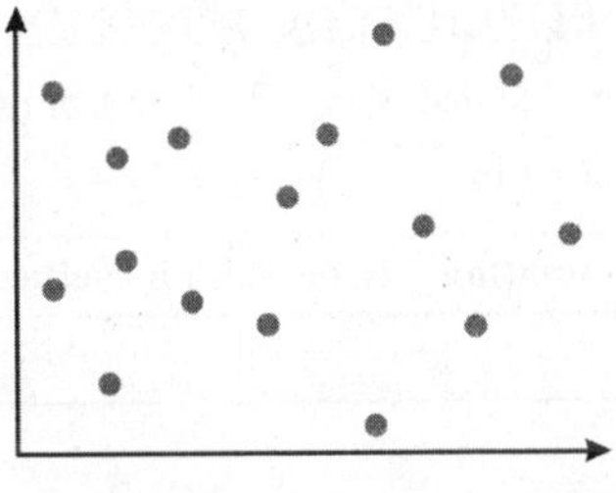

Negative Correlation

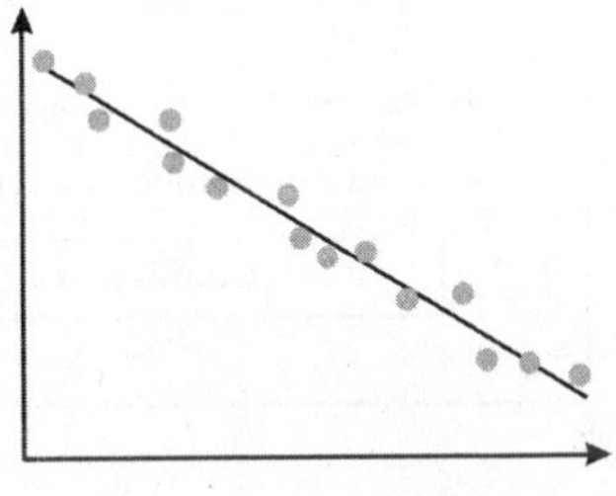

Non-Linear (or) Curvilinear Correlation
• Values of both quantitative variables increase together up to certain point (like positive relationship) and then as one increases, other decreases (negative relationship) or vice versa.
• **Examples:** Time after food intake and Post-prandial Blood Sugar level in OGTT.

Correlation Co-efficient and Degree of Association

r-value	Degree of association
0.0	No association
± 0.1 to ± 0.2	Negligible
± 0.2 to ± 0.4	Weak
± 0.4 to ± 0.7	Moderate
± 0.7 to ± 1.0	Strong
± 1.0	Perfect

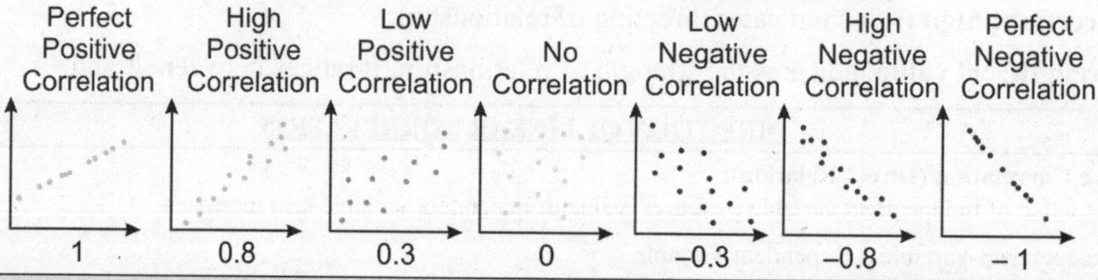

Pearson's Correlation Coefficient

- It is used in ungrouped series
- It is used when associated variables are normally distributed
- Symbol is 'r'

Spearman's Rank Correlation Coefficient

- It is used in grouped series
- It is used when associated variables are not normally distributed
- Symbol is 'rho (r)'

Multiple Correlation Coefficient

- It is used for calculation of correlation between one variable (dependent) and combination of two or more variables (independents)

COEFFICIENT OF DETERMINATION

- It is percentage of variation in variable that is explained by 1 or more of others
- Generally obtained in a regression setup

Coefficient of determination = (Correlation coefficient) 2 = r2. (0 < r2 < + 1)

REGRESSION

- Measure of change in measurement of one variable (response or outcome or dependent variable y') when there is a unit change in another variable (explanatory or predictor or independent variable 'x').
- Regression analysis is a statistical method to reduce effect of confounders on dependent variable in a research study.

SIMPLE REGRESSION ANALYSIS

It would estimate variation observed in one dependent variable for every unit change in one independent variable.

$$y = \alpha + \beta x$$

where, α is a constant and β is the regression co-efficient; y = dependent variable, x = independent variable

MULTIPLE REGRESSION ANALYSIS

It would estimate the variation observed in one dependent variable for every unit change in more than one independent variable (x1, x2, x3, x4 .....etc).

$$y = \alpha + \beta 1\, x1 + \beta 2\, x3 + \beta 3\, x4$$

α is a constant and β1, β2, β3 are the regression co-efficients; y = dependent variable, x1,x2, x3 = independent variables

- **Regression coefficient**
  It is average change in dependent variable for each unit of independent variable.
- **Regression line**
  Approximation of relationship between independent variable and dependent variable.

TYPES OF REGRESSIONS

- **Simple linear regression:** Only one dependent variable and one independent variable.
- **Multiple linear regression:** Only one dependent variable and more than one independent variable.
- **Simple curvilinear regression:** Only one dependent variable and one independent variable, with some power of independent variable.
- **Multiple curvilinear regression:** Only one dependent variable and more than one independent variables, with some power of independent variables).

Types of Regression analysis:

Type of Regression analysis	Dependent variable	Independent variable
Simple Linear Regression	One Dependent Quantitative variable	One independent Quantitative variable
Multiple Linear Regression	Quantitative variable	More than One Quantitative variable
Simple Logistic Regression	One Dependent Quantitative variable (dichotomous)	Qualitative/Quantitative/Mixed
Multiple Logistic Regression	Quantitative variable	Qualitative/Quantitative
ANOVA	Quantitative	Qualitative
ANOVA	Quantitative	Qualitative + Quantitative
ANOVA	Set of Quantitative	Set of Quantitative

MULTIPLE CHOICE QUESTIONS

1. Correlation between variables A and B in a study was found to be 1.1. This indicates: [NEET Pattern 2016]

(a) Very strong correlation
(b) Moderately strong correlation
(c) Weak correlation
(d) Computational mistake in calculating correlation

Ans: (d)

- Correlation can never be greater than 1.

2. A lecturer states that correlation coefficient between prefrontal blood flow under cognitive load and severity of psychotic symptoms in schizophrenic patients is 1.24. You can therefore conclude that: [AIIMS 2010]

(a) Prefrontal blood flow under cognitive load is a good predictor of the severity of psychotic symptoms in schizophrenic patients
(b) Prefrontal blood flow under cognitive load accounts for a large proportion of the variance in psychotic symptoms in schizophrenic patients
(c) Psychosis or schizophrenia is in some way a cause or partial cause of low prefrontal blood flow under cognitive load
(d) Lecturer has reported the correlation coefficient incorrectly

Ans: (d)

- Correlation can never be greater than 1.

3. A cardiologist found a highly significant correlation coefficient ($r = 0.90$, $p = 0.01$) between the systolic blood pressure values and serum cholesterol values of patients attending his clinic. Which of the following statements is a wrong interpretation of correlation coefficient observed? [NEET Pattern 2015]

(a) Since there is a high correlation, the magnitudes of both the measurements are likely to be close to each other
(b) Patient with high level of systolic BP is also likely to have high level of serum cholesterol
(c) Patient with a low level of systolic BP is also likely to have low level of serum cholesterol
(d) About 80% of variation in systolic blood pressure among his patients can be explained by their serum cholesterol values and vice versa

Ans: (a)

- **$r = 0.90$ indicates that there is positive correlation**. Patient with high level of systolic BP is also likely to have high level of serum cholesterol. Patient with a low level of systolic BP is also likely to have low level of serum cholesterol.
- **p is 0.01 which indicates that it is significant value.**

- **Square of r is COEFFICIENT OF DETERMINATION= 81=** About 80% of variation in systolic blood pressure among his patients can be explained by their serum cholesterol values and vice versa.
- Since there is a high correlation, the magnitudes of both the measurements are likely to be close to each other–**Not true.**

4. Which of the following is not true about 'correlation'? [AIIMS 2007]

(a) It indicates degree of association between two characteristics

(b) Correlation coefficient of 1 means that the two variables exhibit linear relationship

(c) Correlation can measure risk

(d) Causation implies correlation

Ans: (c)

- Correlation does not imply causation.
- Causality can be established by 'Hill's Criteria'.

5. Best way to study relationship between two variables is: [NEET Pattern 2012]

(a) Bar chart (b) Scatter diagram

(c) Histogram (d) Pie chart

Ans: (b)

- Refer previous chapter.

6. If we know the value of one variable in an individual and wish to know value of another variable, we calculate: [AIIMS 2017]

(a) Coefficient of correlation (b) Coefficient of regression

(c) SE of mean (d) Geometric mean

Ans: (b)

- Regression provides structure of relationship between 2 Quantitative variables–**Quantification.**

7. If correlation of height with age is given by equation y = a + biopsy, what would be nature of graph? [NEET Pattern 2015]

(a) Straight line (b) Parabola

(c) Hyperbola (d) Sigmoid curve

Ans: (a)

- This is a Simple linear regression Equation: **<u>y = a + bx.</u>**
- Nature of the graph will be a straight line.

8. What can be true regarding coefficient of correlation between IMR and economic status? [AIIMS 2010]

(a) $r = +1$ (b) $r = 0$

(c) $r = +0.22$ (d) $r = -0.8$

Ans: (d)

- Correlation between IMR and economic status has to be inversely proportional
- So increase in economic status results in reduced IMR and vice versa
- Thus it is an example perfect negative correlation.

9. Correlation Coefficient between Smoking and Lung Cancer was found to be 1.4. This indicates: [AIIMS 2007]

(a) Weak correlation (b) Moderate correlation
(c) Strong correlation (d) Mistake in calculation

Ans: (d)

- Correlation can never be greater than 1.

10. Study finds correlation coefficient of + 0.7 between self reported work satisfaction and expectancy of life in random sample of 5000 corporate workers (p = 0.01). This means that: [AIIMS 2017]

(a) Work satisfaction improves life expectancy
(b) Strong statistically significant positive association between work satisfaction and life expectancy
(c) 70% people who enjoy work shall live longer
(d) 70% association between work satisfaction and life expectancy

Ans: (b)

- p Value is 0.01–Significant.
- **Correlation coefficient of + 0.7 indicates positive correlation.**
- Strong statistically significant positive association between work satisfaction and life expectancy exists.

11. Total Cholesterol level = a + b (calorie intake) + c (physical activity) + d (body mass index); is an example of: [NEET Pattern 2005]

(a) Simple linear regression (b) Simple curvilinear regression
(c) Multiple linear regression (d) Multiple logistic regression

Ans: (c)

- **No need to explain. This is multiple linear regression.**

12. A coefficient of correlation value of "r = +0.8" indicates: [DNB 2005]

(a) Strong direct relationship between two variables
(b) Strong inverse relationship between two variables
(c) Insignificant association between two variables
(d) One variable is the cause of the other variable

Ans: (a)

- r = +0.8 indicates positive correlation and it is closer to +1.
- So, it is strong correlation between 2 variables.

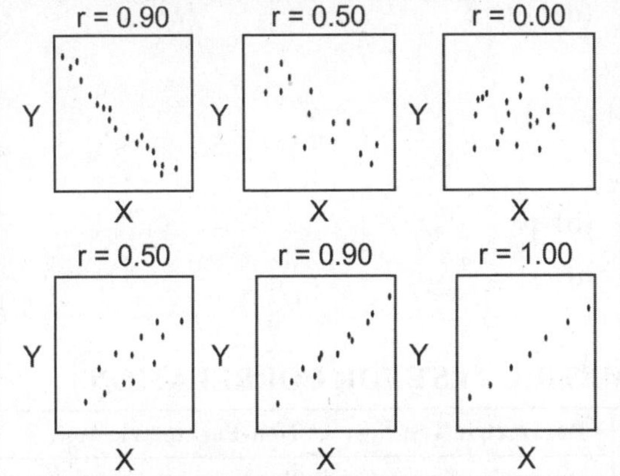

Types of Correlation

- Perfect positive correlation
- Perfect negative correlation
- Partial positive correlation
- Partial negative correlation
- Absolutely no correlation

13. Correlation in height and weight are measured by: **[NEET Pattern 2013]**

(a) Coefficient of variation (b) Range of variation

(c) Correlation coefficient (d) None

Ans: (c)

- Correlation coefficient deals with **STRENGTH** of relationship between 2 variables.
- Regression provides structure of relationship between 2 quantitative variables.

14. Mosquitoes decrease as height increases in: **[NEET Pattern 2013]**

(a) Positive correlation (b) Negative correlation

(c) Bidirectional (d) Zero correlation

Ans: (b)

- <u>This is an example of negative correlation.</u>

15. Strong correlation is signified by a correlation coefficient of: **[DNB 2011]**

(a) Zero (b) 1

(c) Less than 1 (d) More than 1

Ans: (b)

- Correlation can never be greater than 1.
- Strongest correlation is seen with +1.

16. If r = 2.86, it means: **[DNB 2009]**

(a) Positive correlation (b) Negative correlation

(c) No correlation (d) It is a wrong statement

Ans: (d)

- Correlation can never be greater than 1.

17. Correlation co-efficient varies between: **[NEET Pattern 2014]**

(a) 0 to + 1
(b) 0 to –1
(c) –1 to + 1
(d) 1 to 2

Ans: (c)

- Correlation co-efficient varies between –1 to + 1.

18. Pearson correlation is denoted by:

(a) r
(b) p
(c) c
(d) s

Ans: (a)

PARAMETRIC AND NON-PARAMETRIC TEST FOR CORRELATION

Condition	Dependent Variable	Parametric Test	Non-Parametric Test
Correlation between 2 quantitative variables	Continuous/Discrete	Pearson's correlation coefficient (r)	Spearman's correlation co-efficient (ρ)

19. Regression equation is also called as:

(a) Line of average equation
(b) Estimating equation
(c) Prediction equation
(d) All of the above

Ans: (d)

20. If ordinal data is given then which of the following correlation may be used?

(a) Spearman correlation
(b) Pearson correlation
(c) Point correlation
(d) No correlation

Ans: (a)

PARAMETRIC AND NON-PARAMETRIC TEST FOR CORRELATION

Condition	Dependent Variable	Parametric Test	Non-Parametric Test
Correlation between 2 quantitative variables	Continuous/Discrete	Pearson's correlation coefficient (r)	Spearman's correlation co-efficient (ρ)

21. In equation y = a + bx, then x is:

(a) Slope of line
(b) Intercept of line
(c) Direction of correlation
(d) Any of the above

Ans: (a)

- a is a constant.
- b is regression co-efficient.
- y = dependent variable.
- x = independent variable. Indicates slope of the line.

22. Degree of association between 2 variables is given by:

(a) Correlation (b) Regression

(c) Causation (d) Z test

Ans: (a)

- Correlation coefficient deals with **STRENGTH** of relationship between 2 variables.
- Regression provides structure of relationship between 2 quantitative variables.

23. If we know the value of one variable and wish to know value of another variable then it is calculated by:

(a) Correlation coefficient (b) Regression coefficient

(c) Standard error of mean (d) Geometric mean

Ans: (b)

- Regression provides structure of relationship between 2 quantitative variables.

NUTSHELL

- 'r' value between 0 and – 1 indicates negative relationship while 'r' value between 0 and + 1 indicates positive relationship between 2 variable
- r = 0 indicates 'No Correlation' between the 2 variables
- **r = Correlation Coefficient = –1 < 0 < +1**

PARAMETRIC AND NON-PARAMETRIC TEST FOR CORRELATION

Condition	Dependent Variable	Parametric Test	Non-Parametric Test
Correlation between 2 quantitative variables	Continuous/Discrete	Pearson's correlation coefficient (r)	Spearman's correlation coefficient (ρ)

CHAPTER 15 SURVIVAL ANALYSIS

- Deals with research studies on how long people with life threatening disease would survive following a treatment.
- Analytical methods are used to draw inferences regarding the chances of:
  - Surviving or Dying.
  - Getting cured or getting diseased.

  over various points of time during the study period are answered by "survival analysis".

LIFE TABLE METHOD:

- Life table method for analyzing a survival data accounts for differences in the follow-up time and also account for changes in survival rate over time.
- It breaks the follow-up period of the study into intervals and then calculates the proportion of patients at risk at the start of each interval who would survive until the start of the next interval.

KAPLAN-MEIER METHOD:

- Kaplan-Meier method of life table analysis of survival data uses the occurrence of next death among the patients followed up, to define the end of the last class interval and to start the new class interval.
- Survival curves are drawn to show the percentage of patients surviving versus time.
- It can be used to describe the survival data of a group of patients using a survival curve.
- It can also be used to compare the survival curves of two independent groups. E.g., Difference between median survival times of two treatment groups.

LOG-RANK TEST:

- In Kaplan-Meier method, a **non-parametric test** is applied to test the null hypothesis, that there is no difference in median survival rates of the patients treated with two different treatment modalities (comparing two survival curves) called as ***Log-Rank Test.***
- Log-Rank test **does not control the confounding factors** present in a study.

COX PROPORTIONAL HAZARDS REGRESSION MODEL:

- When the dependent or outcome variable is survival data (i.e., in case of a time-to-event data), the regression method applied for analysis is called as Cox Regression model.
- It is also called as Proportional Hazard regression as it makes use of the hazard function.

- It is used to compare the survival curves of ≥ 2 **independent groups**.
- This test **controls confounding factors** present in the study.

CONDITIONAL PROPORTIONAL HAZARDS REGRESSION MODEL:

- It is used to compare the survival curves of ≥ 2 **paired groups**. E.g., Difference between the median survival time of patients before and after treatment.

MULTIPLE CHOICE QUESTIONS

1. Q-test is used for detecting: **[NEET Pattern 2013]**

(a) Outliers
(b) Interquartile range
(c) Difference of means
(d) Difference of proportions

Ans: (a)

DIXON'S Q-TEST:

- **Use:** ***To detect and eliminate outlier from a distribution.***

ASSUMPTIONS:

- Data has to be normal distributed.
- Not supposed to be used more than once in the same data set.
- **Method:** Measure the difference between a suspect value and next closest value, and then compare it with total range of observations.
- Alternative method to detect outliers: **Grubbs' Test**.

2. Test is used to compare Kaplan-Meier survival curve: **[NEET Pattern 2012]**

(a) ANOVA
(b) Bland Altmann analysis
(c) Chi-square test
(d) Cox proportional hazards test

Ans: (d)

- Cox proportional hazards test is used to compare the survival curves of ≥ **2 independent groups.**
- Bland Altmann analysis comparison of existing test with a gold standard test.
- Chi-square test is non parametric test.
- ANOVA also is a test of significance.

3. Q-test is used for detecting: **[NEET SS Pattern 2017]**

(a) To detect outliers
(b) Interquartile range
(c) Difference between paired means
(d) Difference between independent proportions

Ans: (a)

- Detects and eliminates outlier from a distribution.

4. Which of the following test is used for comparison between survival rates of 2 treatment groups?

(a) Logistic regression
(b) Log-Rank test

(c) Kendall rank correlation (d) Spearman rank correlation

Ans: (b)

- In Kaplan-Meier method, a **non-parametric test** is applied to test the null hypothesis, that there is no difference in median survival rates of the patients treated with two different treatment modalities (comparing two survival curves) called as ***Log-Rank Test.***

5. A Gastro-surgeon wants to compare the survival data of colon cancer patients following three different surgical procedures. Which is the most appropriate statistical test for comparing survival data of these 3 groups of patients?

(a) Log-Rank Test
(b) Kruskal Wallis test
(c) Conditional proportional hazards regression model
(d) Cox proportional hazards regression model

Ans: (d)

Condition	Survival Analysis Method
Description of survival data of one group	Kaplan-Meier survival curve
Comparison of survival data of 2 independent groups	Log-Rank Test
Comparison of survival data of ≥ 2 paired groups	Conditional proportional hazards regression model
Comparison of survival data of ≥ 2 independent groups	Cox proportional hazards regression model

NUTSHELL

Survival or life table analysis:

- It is used to measure a defined outcome in a population (or sample), followed over a variable time period.
- Subjects are entered independently and observed at different time periods and for varying length of time.
- Subjects lost to follow up are included in analysis for the time period they were observed.
- Contribution of each subject is weighted by the time under observation (measured in person years).
- **Types:**
  - Demographic or population life table → based on application of national or other age specific mortality data to a theoretical population.
  - Cohort or clinical life table → based on data obtained from an observed or studied sample. Each subject has a known initial point and a well defined endpoint of observation with dichotomous outcome (Alive or dead).

Kaplan Meier Method **of analysis of clinical Cohort** → provides nonparametric estimate of cumulative incidence for defined outcomes based on incomplete observations.

Cronbach's α (alpha) is a measure of *internal consistency or reliability.* Alpha varies from 0 to 1.

Cronbach's α	≥ 0.9	0.9 > α ≥ 0.8	0.8 > α ≥ 0.7	0.7 > α ≥ 0.6	0.6 > α ≥ 0.5	α < 0.5
Internal Consistency	Excellent	Good	Acceptable	Questionable	Poor	Unacceptable

CHAPTER 1
BASICS OF SCREENING

IMPORTANT TERMS
EFFICIENCY: • Inputs in terms of expenses and time result in reasonable amount of outputs in terms of improved health and satisfaction. **SCREENING:** • Testing for infection or disease in populations or in individuals don't seek healthcare. • **Example:** Serological testing for AIDS virus in blood donors, neonatal screening, premarital screening for syphilis. **CASE FINDING:** • Use of clinical and/or laboratory tests to detect disease in individuals seeking healthcare for other reasons **Example:** VDRL test to detect syphilis in pregnant women. **DIAGNOSTIC TESTS:** • Use of clinical and/or laboratory procedures to confirm or refute the existence of disease or true abnormality in patients with signs and symptoms presumed to be caused by the disease. • **Example:** VDRL testing of patients with lesions suggestive of secondary syphilis.

SCREENING – It is active search for unrecognized disease or defect by means of rapidly applied tests, examinations or other procedures in apparently healthy individuals.

Aim–To bring those who are "apparently abnormal" under medical supervision and treatment.

CONCEPT OF SCREENING
• Active search for disease among the apparently normal • Early detection of 'Hidden' Disease (Secondary Prevention) • Capable of wide application • Relatively inexpensive • Requires little Physician-time

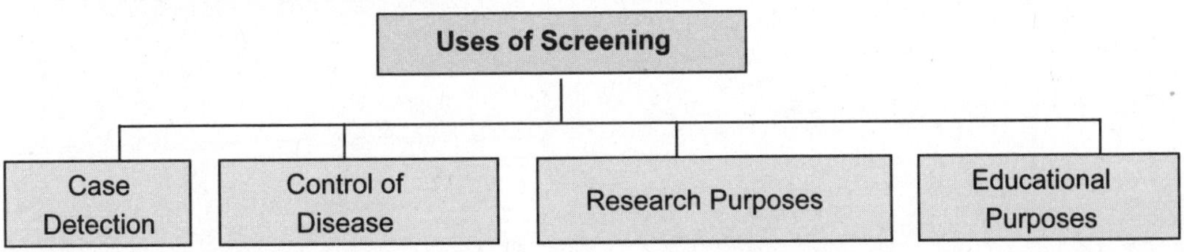

DIFFERENCE BETWEEN SCREENING AND DIAGNOSTIC TESTS	
SCREENING TEST	**DIAGNOSTIC TEST**
Done on apparently healthy persons	Done on those with indications or sickness
Applied to groups	Applied to individual patients, all possible diseases are considered
Test results are arbitrary and final	Diagnosis is not final. It can be modified with new evidence
Based on one criterion or cut-off point	Based on evaluation of a number of symptoms, signs and laboratory findings
Less accurate	More accurate
Less expensive	More expensive
Not a basis for treatment	Used as a basis for treatment
Initiative come from the investigator	Initiative comes from a patient with a complaint
The test must be–Reliable, Valid, Efficient, Practical and give good yield	The test must be–Reliable, Valid and Efficient

CHAPTER 2

CRITERIA OF SCREENING

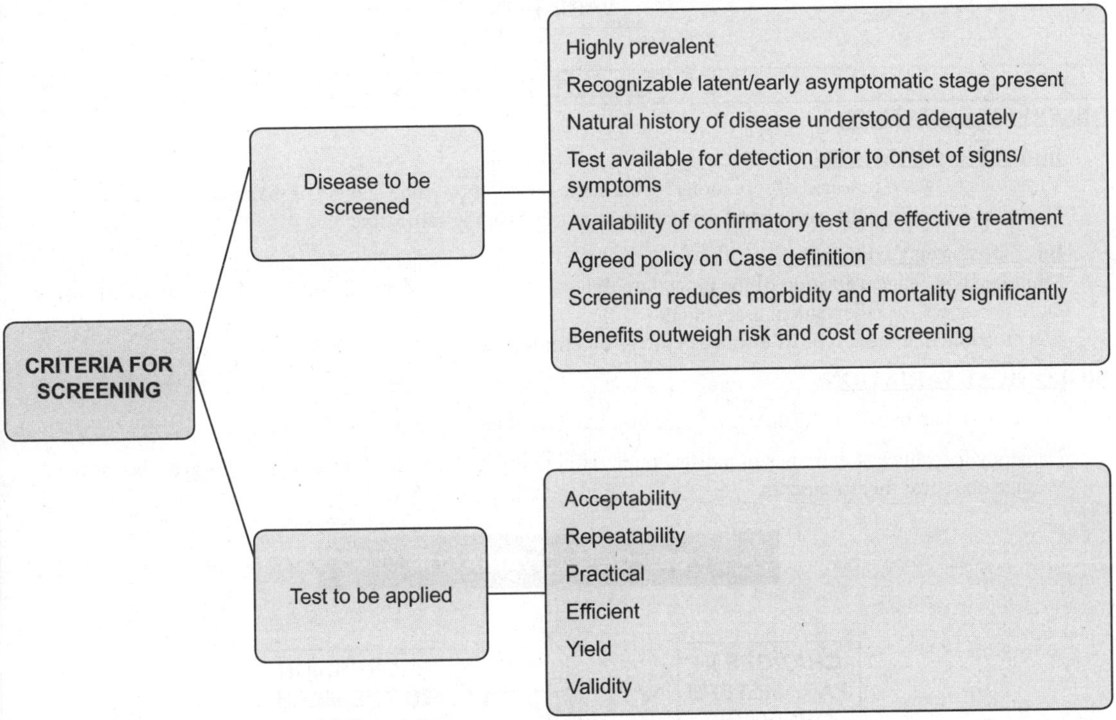

TEST CRITERIA

i. ACCEPTABILITY:

- **Simple and Safe** (Painful, discomforting or embarrassing are not likely to be acceptable to the population).
- **Example:** Per Rectal examination is not acceptable in screening camps.

ii. REPEATABILITY (RELIABILITY, PRECISION, REPRODUCIBILITY):

- Test must give consistent results when repeated more than once on same or material, under the same conditions.

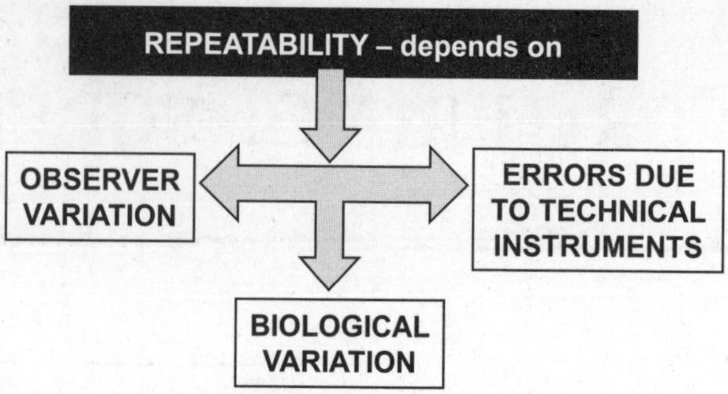

REPEATABILITY

OBSERVER VARIATION:

- **Intra Observer Variation:**
  Variations between repeated observations by the same observer on same subject or material at same time
  Minimized by taking the **average of several** measurements at the same time
- **Inter Observer Variation:**
  Minimized by Standardization of the procedure, Intensive training of the individuals and making use of two or more observers for independent assessment
  It is probable that these type of errors can **never be eliminated**

BIOLOGICAL VARIATION:

- Associated with many of physiological variables like BP, blood sugar, serum cholesterol etc.
- Tendency for values at extremes of a distribution, either vary high or very low to regress towards the mean or average on repeat measurements.

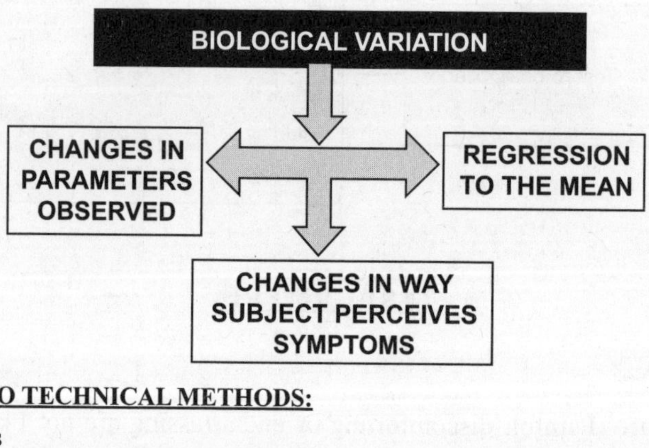

ERRORS RELATING TO TECHNICAL METHODS:

- Defective instruments
- Erroneous calibration
- Faulty reagents
- Test itself might be inappropriate or unreliable

> Observer variation–Checked by repeat measurements at the same time
> Biological variation–Tested by repeated measurements over the period of time

iii. VALIDITY (ACCURACY):

- Refers to what extent the test accurately measures what it has to measure.
- Ability to distinguish those who are diseased from the healthy.
- Closeness of the 'measured' values to the 'true' values.
- Sensitivity and specificity are 2 components of validity.

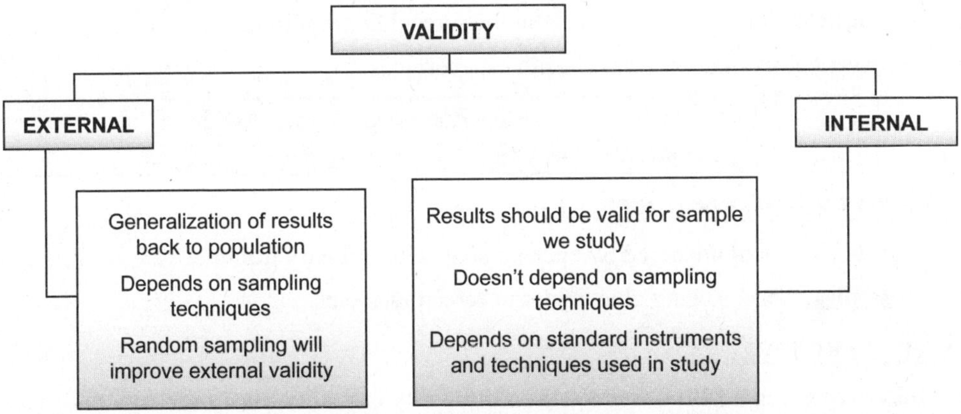

DIFFERENCES BETWEEN PRECISION AND ACCURACY

Precision	Accuracy
Degree to which a variable has nearly the same value when measured several times Comparison among repeated measures	Degree to which a variable actually represents what it is supposed to represent Comparison with a reference standard

Range chart R – chart	Mean chart Levy Jennings (LJ) chart Shewhart control chart

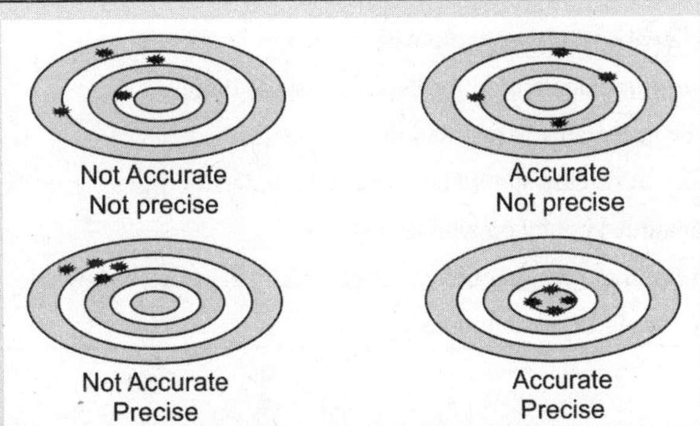

Predictive value of a positive test:

Indicates the probability that a patient with positive test result has the disease.

Predictive value of a positive result falls as disease prevalence declines.

iv. **YIELD:**

- Amount of previously unrecognized disease that is diagnosed as a result of screening.
- **Depends on:**
  - Prevalence of the disease screened (High Risk Group = ↑ yield)
  - Sensitivity and Specificity of the Screening test
  - Participation of the individuals in the screening programme
  - Availability of appropriate health care services

"Yield α Prevalence of the Disease"
High risk populations are usually selected for screening, thus increasing yield.

v. **EASY TO EXECUTE/PRACTICAL:**

- Test should be administered by even persons with ordinary training.
- **Examples:** ANM's using glucometer to screen Diabetes.

vi. **COST EFFECTIVE:**

- Improvement in health or reduction in morbidity/mortality must outweigh resources and time spent.

Screening test must satisfy the criteria of not only acceptability, repeatability and validity but also **yield, simplicity, safety, rapidity, cost effectiveness**

PRINCIPLES OF SCREENING (WHO) **Suitability of a Disease for Screening (Criteria)**
• Disease should be an important health problem
• Should be an effective treatment available for the disease
• Facilities for diagnosis and treatment should be available
• Should be a latent or early asymptomatic stage of the disease
• Should be a test or examination for the diagnosis of disease
• Test should be acceptable to population
• Natural history of disease should be adequately understood
• Should be an agreed policy on who to treat
• Total cost of finding a case should be economically balanced as a whole
• Case-finding should be a continuous process

CHAPTER 3

TYPES OF SCREENING

MASS SCREENING:

- Screening of a whole population or a subgroup.
- Useless unless backed up by suitable treatment to reduce duration of illness or alter its final outcome.

HIGH RISK/SELECTIVE SCREENING:

- Screening is applied selectively to high-risk groups.
- It is effective and economical–**most productive.**
- **Example:** Screening for cancer cervix in the lower social groups.

MULTIPHASIC SCREENING:

- Application of two or more screening tests in combination to a large number of people at one time than to carry out separate screening tests for single diseases.

MULTIPURPOSE SCREENING:

- Screening of a group of population by application of 2 or more tests, at one time to detect more number of diseases.
- **Example:** Pregnant women is tested for Hb%, VDRL, ELISA for HIV.

OPPORTUNISTIC SCREENING/CASE FINDING SCREENING:

- Screening of a patient who consults the doctor for some other purpose.

	Prescriptive screening	**Prospective screening**
Definition	People screened for own's benefit	People screened for other's benefit
Essential purpose	Case detection	Disease control
Request for screening	No specific request	Specific request from authority
Example(s)	Neonatal screening Pap smear Urine for sugar	Screening of immigrants HIV screening among Sex workers

CHAPTER 4

SCREENING TESTS

SCREENING TESTS
• Sensitivity • Specificity • Positive predictive value • Negative predictive value • Percentage of false +ve • Percentage of false –ve

SENSITIVITY:

- Ability of a test to identify correctly all those who have the disease–True Positive (TP).
- Statistical index of diagnostic accuracy.

$$\textit{Sensitivity} = \frac{\text{True positive [TP]}}{\text{True positive [TP] + False negative [FN]}} \times 100 \text{ or } \frac{\text{True positive [TP]}}{\text{Total diseased}} \times 100$$

SPECIFICITY:

- Ability of a test to identify correctly all those who do not have the disease (healthy)–True Negative (TN).

$$\textit{Specificity} = \frac{\text{True negative [TN]}}{\text{True negative [TN] + False positive [FP]}} \times 100 \text{ or } \frac{\text{True negative [TN]}}{\text{Total not diseased}} \times 100$$

EXTRA EDGE

- If Sensitivity increases if TP increases and FN decreases
- If Specificity increases if TN increases and FP decreases
- ***Sensitivity and specificity are inversely related***
- False +ve error is more serious than false –ve error
- For screening tests sensitivity is important whereas for diagnostic tests specificity is very important
- **High false +ve** → Low specificity, Low prevalence, High specificity

INHERENT PROPERTIES OF A SCREENING TEST

Sensitivity, Specificity and Predictive accuracy.

FALSE NEGATIVE:

- It means patients who actually have the disease are told that they do not have the disease.
- A screening test which is very sensitive has few "false negatives".
- Patient ignores signs / symptoms → postponement of treatment → ↑ morbidity and mortality

$$\text{False Negative Rate} = \frac{c}{a + c} \times 100 = \frac{FN}{TP + FN} \times 100$$

False negative rate (β) = 1 – sensitivity

Power or a test (sensitivity) = (1 – β) = 1 – False negative rate

FALSE POSITIVE:

- It means patients who do not have the disease are told that they have.
- **A screening test with a high specificity will have few false positives.**
- Causes burden to diagnostic facility; discredit to the screening programme
- A test which has very high sensitivity (though at the cost of low specificity) will have high FP.

$$\text{False Positive Rate} = \frac{b}{b + b} \times 100 = \frac{FP}{FP + TN} \times 100$$

False Positive Rate (α) = 1 – sensitivity

POSITIVE PREDICTIVE VALUE (PPV):

- PPV is the probability that a person with positive test result truly have disease.
- Proportion of patients with positive test results who are correctly diagnosed.

$$PPV = \frac{a}{a + b} \times 100 = \frac{TP}{TP + FP} \times 100$$

PPV α Prevalence; Specificity; Sensitivity

- If sensitivity and specificity of a test are constant, predictive values are largely dependent on prevalence of disease in the population in which diagnostic test is being applied.

NEGATIVE PREDICTIVE VALUE (PPV):

- NPV is the probability that a person with - ve test result does not have disease.
- Proportion of patients with negative test who are correctly diagnosed.

$$NPV = \frac{d}{c + d} \times 100 = \frac{TN}{TN + FN} \times 100$$

- **NPV is inversely related to the Prevalence of the disease in population**

$$\text{Predictive value of positive test}^{Q} = \frac{\text{True positive [TP]}}{\text{True positive [TP]} + \text{False positive [FP]}} \times 100 \text{ or } \frac{\text{True positive}}{\text{Total positive declared by test}} \times 100$$

$$\text{Predictive value of negative test}^{Q} = \frac{\text{True negative [TN]}}{\text{True negative [TN]} + \text{False negative [FN]}} \times 100 \text{ or } \frac{\text{True negative [TN]}}{\text{Total negative declared by test}} \times 100$$

ACCURACY:

- Degree of closeness of a measured or calculated quantity to its true value.
- Predictive accuracy reflects the diagnostic power of the test.

$$\text{Accuracy} = \frac{TP + TN}{TP + FP + FN + TN} \times 100$$

Accuracy = [(sensitivity) × (prevalence)] + [specificity) + (1 – prevalence)]

BAYES THEOREM:

- It is statement of conditional probability of a disease for an individual with a positive screening test.

$$\text{PPV} = \frac{[\text{Sensitivity} \times \text{Prevalence}]}{[\text{Sensitivity} \times \text{Prevalence}] + (1 - \text{Specificity})\,(1 - \text{Prevalence})} \times 100$$

$$\text{NPV} = \frac{[\text{Specificity} \times (1 - \text{Prevalence})]}{[\text{Specificity} \times (1 - \text{Prevalence})] + [(1 - \text{Sensitivity}) \times \text{Prevalence})]} \times 100$$

EXTRA EDGE

SIMULTANEOUS OR PARALLEL SCREENING:

- Multiple tests are used simultaneously
- Individual is considered to have tested

$$\text{Net Sensitivity} = \frac{\text{Total no. of test positives}}{\text{Total no. of diseased subjects}} \times 100$$

$$\text{Net Specificity} = \frac{\text{No. of true negatives in both the tests}}{\text{Total no. of healthy subjects}} \times 100$$

Net sensitivity = (Sensitivity of test 1 + Sensitivity of test 2) – (Sensitivity of test 1 × Sensitivity of test 2)

Net specificity = Specificity of test 1 + Specificity of test 2)

- Positive → if he or she has a positive result on any one or more of the tests.
- Negative → if he or she tests negative on all of the tests.
- Here there is a net gain in sensitivity and a net loss in specificity compared to using either test alone.

SEQUENTIAL OR SERIES SCREENING

- Here we retest those who tested positive on the first test.
- There is a loss in net sensitivity and a gain in net specificity compared with either test alone.

$$\text{Net Sensitivity} = \frac{\text{Final no. of true positives}}{\text{Total no. of diseased subjects}} \times 100$$

$$\text{Net Specificity} = \frac{\text{Total no. of true negatives}}{\text{Total no. of healthy subjects}} \times 100$$

Net sensitivity = Sensitivity of test 1 × Sensitivity of test 2

Net specificity = (Specificity of test 1 + Specificity of test 2) – (Specificity of test 1 × Specificity of test 2)

NUTSHELL

Net Remarks	Tests in Series (Sequential)	Tests in Parallel (Simultaneous)
Net Sensitivity	Decreases	Increases
Net Specificity	Increases	Decreases
Net PPV	Increases	Decreases
Net NPV	Decreases	Increases

CHAPTER 5

PROBABILITY AND ITS RULES

- Post-test probability of a disease in a population (PTP) is same as PPV.
- Pre-test probability of a disease in a population (pTP) is same as Prevalence.

LIKELIHOOD RATIO:

- Measure of diagnostic accuracy of a test
- Likelihood – probability, possibility, chance, odds
- Likelihood ratio – ratio of probabilities

$$LR = \frac{\text{Probability of an individual with the disease having the test result}}{\text{Probability of an individual without the disease having the test result}}$$

Likelihood ratios combines elements of test sensitivity and test specificity.

$$\text{Likelihood ratio positive} = \frac{\text{Sensitivity}}{(1 - \text{specificity})}$$

$$\text{Likelihood ratio negative} = \frac{(1 - \text{specificity})}{\text{Specificity}}$$

- LR + > 1 indicates an increased probability that target disease is present after (positive) test
- LR – < 1 indicates a decreased probability that target disease is present after (negative) test
- Smaller the ratio LR +, better the test will be

Qualitative Strength	Likelihood Ratio +	Likelihood Ratio –
Excellent	10	0.1
Very good	6	0.2
Fair	2	0.5
Useless	1	1

Interpreting Likelihood Ratio:

LR values for a test		Interpretation
Positive	**Negative**	**Changes in rate after test**
> 10	< 0.1	Conclusive
5 - 10	0.1 - 0.2	Moderate
2 - 5	0.5 - 0.2	Small
1 - 2	0.5 - 1	Negligible

PRE TEST ODDS AND POST TEST ODDS

- Post test odds = Chances that a patient has a disease after the test
- Pre test odds = Chances that a patient has a disease before the test

Post test odds = Pre test odds × Likelihood Ratio

CHAPTER 6

FORMULAS

BASIC MODEL OF 2 × 2 CONTINGENCY TABLE

Screening test results	Disease condition in the population		Total
	Diseased	Not diseased	
Positive	a (True Positives)	b (False Positives)	a + b
Negative	c (False Negatives)	d (True Negatives)	c + d
Total	a + c	b + d	a + b + c + d

- Sensitivity = a/(a+c) × 100
- Specificity = d/(b+d) × 100
- Predictive value of positive test = a/(a+b) × 100
- Predictive value of negative test = d/(c+d) × 100
- Percentage of false negatives = c/(a+c) × 100
- Percentage of false positives = b/(b+d) × 100
- Prevalence of the disease = (a+c)/(a+b+c+d) × 100
- Positivity rate or prevalence of positivity = Total positives / Total tested × 100
- Negativity rate or prevalence of negativity = Total negatives / Total tested × 100
- Rate of Disagreement = (False positives + False negatives) / Total tested × 100
- Rate of Agreement = (True positives + True negatives) / Total tested × 100

CHAPTER 7

MISCELLANEOUS

RECEIVER OPERATOR CHARACTERISTIC (ROC) CURVE

- ROC allows knowing optimal specification of test sensitivity and specificity.
- ROC curve plots (Sensitivity) on Y axis **Vs** (1 – specificity) on X-axis.
- Helps to determine the cut-off for a positive test in a screening / diagnostic test.
- Helps to compare two or more screening / diagnostic tests of a disease with each other.

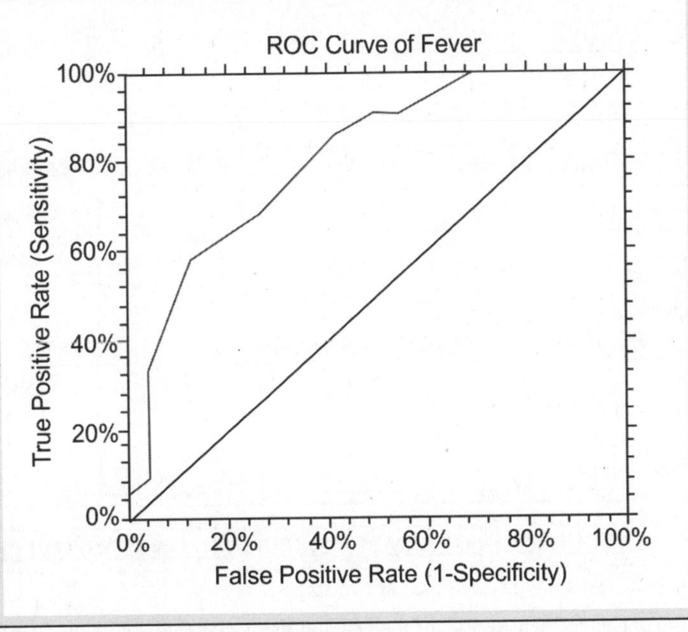

EVALUATION OF SCREENING PROGRAMMES

RANDOMIZED CONTROLLED TRIALS:

- **Ideal.**
- **Limitation**–Cost and logistics are often prohibitive if disease has a low frequency in the population and a long incubation period (e.g., cancer).

Other methods:

- Case control studies and comparison in trends between areas with different degrees of screening coverage.

Concept of Cut Off

- This is valuable for testing continuous variables

$$\text{Cut} - \text{off } \alpha \ \frac{\text{Specificity}}{\text{Sensitivity}} \ \alpha \ \frac{1}{\text{Prevalence}}$$

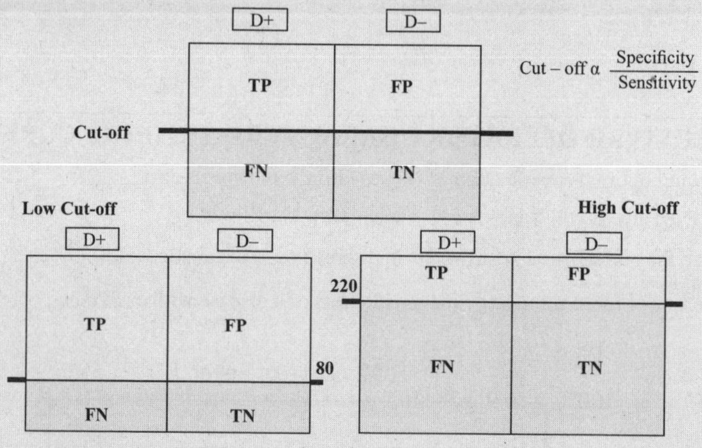

Disease character	Cut-off	Specificity	Sensitivity
High Prevalence eg. Anemia	Low	Low	High
Very lethal disease Early detection improve the prognosis Tolerable false positive Eg. CA prostate, CA Thyroid	Low	Low	High
Chronic disease Early detection doesn't alter the disease outcome much Need to limit false positive Eg. Diabetes, Hypertension	High	Low	Low

CALCULATION OF SAMPLE SIZE BASED ON SENSITIVITY

Sample size in a diagnostic test study is calculated in two stages :

- **First:** Specify the expected "sensitivity" of the test and the acceptable deviation from this sensitivity on either side of the expected sensitivity.

$$a = \frac{\text{Sensitivity} \times (1 - \text{Sensitivity})}{(\text{Deviation})^2}$$

- **Second**: Actual sample size 'N' is calculated by the formula

$$N = \frac{a}{\text{Prevalence}}$$

CONCEPT OF LEAD TIME

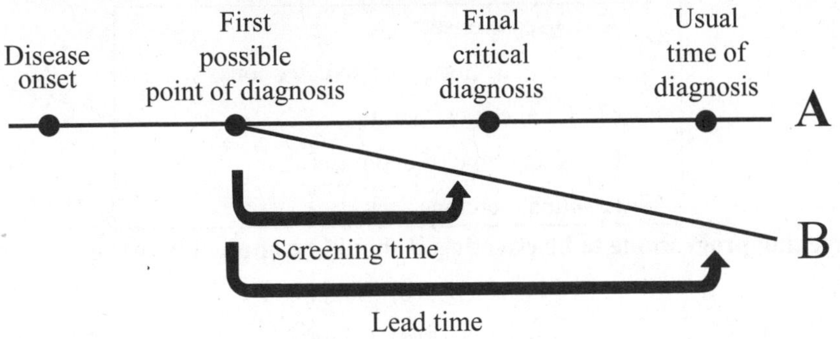

- "Lead time" is the advantage gained by screening, i.e., the period between diagnosis by early detection and diagnosis by other means.
- A = Usual outcome of the disease.
- B = Outcome to be expected when disease is detected at the earliest possible moment.
- (B – A) = Benefits of the screening programmes.

"CONFUSION CLEARED HERE"

Most important and High yield concept (Even toppers get confused few times here)

- Difference between earliest/1st possible point of diagnosis and critical point of diagnosis is **Screening time**
- Difference between earliest/1st possible point of diagnosis and usual point of diagnosis is **Lead time**

BIAS IN SCREENING

LEAD TIME BIAS:

- Systematic error of apparent increased survival from detecting disease in an early stage.

LENGTH TIME BIAS:

- Diseases with a long pre-clinical phase are more likely to be detected during screening.

VOLUNTEER BIAS:

- Not everyone will take part in a screening program.
- There are factors that differ between those willing to get tested and those who are not.
- Willingness depends on perceived risk of disease and intelligence of patient.
- **Example:** Breast cancer screening – more positive outcome if only intelligent people participate. More negative outcomes if only high risk patients participate.

MULTIPLE CHOICE QUESTIONS

1. There are several points in the course of a disease process: **[AIIMS 2008]**

A – Disease onset
B – Point of first possible detection
C – Final critical point
D – Usual time of diagnosis
E – Final outcome

For a screening programme to be effective, it should be applied between:

(a) A and B
(b) A and C
(c) B and C
(d) C and D

Ans: (c)

- Difference between earliest/1st possible point of diagnosis and critical point of diagnosis is **Screening time**
- Difference between earliest/1st possible point of diagnosis and usual point of diagnosis is **Lead time**

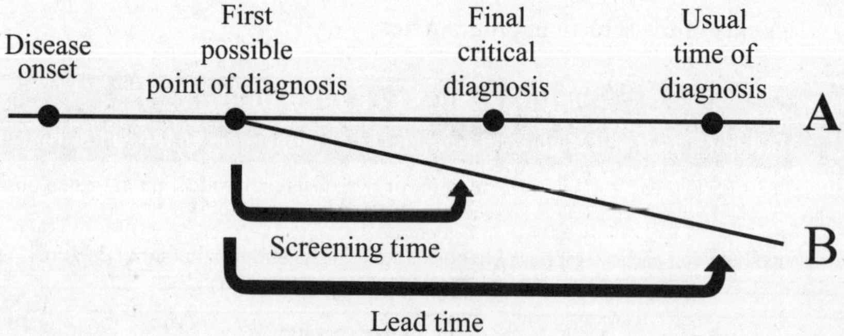

2. Screening is the most commonly used epidemiological tool in school health services. Which level of prevention does it refer to? **[AIIMS 2012]**

(a) Primary
(b) Secondary
(c) Tertiary
(d) Primary and secondary

Ans: (b)

Levels of Prevention:

Level	Phase of disease	Objective	Interventions	Targets
Primordial	Underlying conditions lead to causation	Minimize hazards to health	Inhibit emergence of risk conditions	Total population or selected groups
Primary	Specific causal factors	Reduce incidence	Personal and community efforts	Total population, selected groups, individuals

Contd.

Contd.

Secondary	Early stages of disease	Shorten duration, Reduce prevalence	Early detection and prompt intervention	Individuals with established disease
Tertiary	Late stages of disease	Reduce no./impact of complications	Lessen impact of long-term disease/disability, minimize suffering	Patients

3. Not a part of National Screening Programmes: [NEET Pattern 2014]

(a) Diabetes mellitus (b) Carcinoma cervix

(c) Refractive errors (d) Dental caries

Ans: (d)

- **Refractive errors:** Under 'National Programme for Control of Blindness.
- **Diabetes mellitus:** Under 'National Programme for Prevention and Control of Diabetes, Cardiovascular diseases and Stroke'.
- **Carcinoma cervix:** Under 'National Cancer Control Programme'.

4. Diagnostic power of test to correctly exclude disease is reflected by: [AIIMS 2013]

(a) Sensitivity (b) Specificity

(c) Positive predictivity (d) Negative predictivity

Ans: (d)

- Diagnostic power of screening test to correctly identify a disease: Positive predictive value.
- Diagnostic power of screening test to correctly exclude a disease: Negative predictive value.

5. 'Lead time' refers to the time between: [NEET Pattern 2013]

(a) Disease onset and first critical diagnosis

(b) Disease onset and first possible point of detection

(c) First possible point of detection and final critical point

(d) First possible point of detection and usual time of diagnosis

Ans: (d)

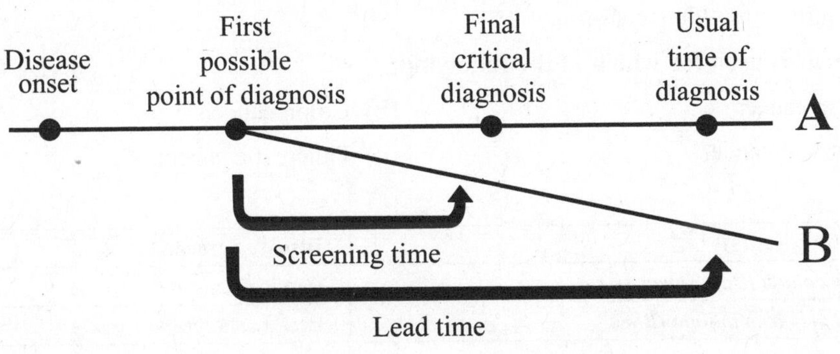

6. **Screening for condition recommended when:** **[NEET Pattern 2012]**
   (a) Low case fatality rate
   (b) Diagnostic tools not available
   (c) No effective treatment available
   (d) Early diagnosis can change disease course because of effective treatment

Ans: (d)

- Screening comes under secondary prevention which deals with early diagnosis and treatment.

7. **Screening of the diseases is which type of prevention?** **[DNB Pattern 2011]**
   (a) Primordial (b) Primary
   (c) Secondary (d) Tertiary

Ans: (c)

Refer above question.

8. **Screening is done because of all *except*:** **[NEET Pattern 2014]**
   (a) Testing for infection or disease in population or in individuals who are not seeking health care
   (b) It is defined presumptive identification of unrecognized disease
   (c) Search for unrecognized disease or defect by means of rapidly applied test, examinations or other procedures in apparently healthy individuals
   (d) Use of clinical/laboratory tests to detect disease in individual seeking health care for other reasons

Ans: (a)

<u>**CONCEPT OF SCREENING:**</u>

- Active search for disease among the apparently normal
- Early detection of 'Hidden' Disease (Secondary Prevention)
- Capable of wide application
- Relatively inexpensive
- Requires little Physician-time

9. **Pap smear is used for which of the following:** **[NEET Pattern 2014]**
   (a) Lung cancer (b) Colon cancer
   (c) Cervical cancer (d) Pancreatic cancer

Ans: (c)

Screening Test(s)	Disease Screened
Papanicolaou (Pap) smear test	Cervical cancer
Breast self examination (BSE)	Breast cancer

Contd.

Contd.

Mammography	Breast cancer
Bimanual oral examination	Oral cancer
ELISA, RAPID, SIMPLE	HIV (National AIDS Control Programme)
Urine for Sugar, Random blood sugar	Diabetes mellitus
AFP (alpha feto-protein)	Developmental anomalies in fetus
Digital rectal examination (DRE)	Prostate cancer
Prostate specific antigen (PSA)	Prostate cancer
Fecal occult blood test	Colorectal cancer

10. Iceberg phenomenon is part of disease, which presents clinically, screening tests are used for: [Bihar 2004]

(a) To detect tip of iceberg
(b) To detect clinical cases
(c) To detect submerged part [clinical cases]
(d) To detect submerged part [sub clinical cases]

Ans: (d)

	Tip of iceberg	**Submerged part of iceberg**
Composition	Clinical cases	Latent, inapparent, presymptomatic, undiagnosed and undiagnosed cases and carriers
Visibility to clinician	Visible	Invisible
Prime importance for	Clinician	Epidemiologist
Detection	Diagnostic tests	Screening tests
Useful level of prevention	Secondary	Secondary

11. Period between the possible time of detection and actual time of diagnosis is: [NEET Pattern 2014]

(a) Lead time
(b) Screening time
(c) Generation time
(d) Serial interval

Ans: (a)

- Difference between earliest/1st possible point of diagnosis and critical point of diagnosis is **Screening time.**
- Difference between earliest/1st possible point of diagnosis and usual point of diagnosis is **Lead time.**

12. Which of the following is an example of Prospective screening? [AIIMS 2011]

(a) Cervical Pap smear in a 40-year-old female
(b) Neonatal screening in a newborn for Hypothyroidism
(c) Screening of immigrants in a country
(d) Urine for sugar screening in a 40-year-old male

Ans: (c)

	Prescriptive screening	Prospective screening
Definition	People screened for own's benefit	People screened for other's benefit
Essential purpose	Case detection	Disease control
Request for screening	No specific request	Specific request from authority
Example(s)	Neonatal screening Pap smear Urine for sugar	Screening of immigrants HIV screening among sex workers

13. Which of the following is not useful as a screening method? **[AIIMS 2013]**

(a) Pap smear for Cervical cancer

(b) CA-125 for Ovarian cancer

(c) Office endometrial washing for Endometrial cancer

(d) USG in Endometrial cancer

Ans: (b)

- **Refer above–question 9.**

14. More sensitive test show: **[NEET Pattern 2017]**

(a) Increased False +ve (b) Decreased False +ve

(c) Increased False –ve (d) Decreased False –ve

Ans: (d)

- Sensitivity represents number of true positives.
- Therefore if it is Increased then there is fall in false –ve.

15. True positivity is indicated by: **[RJ 2006]**

(a) Sensitivity (b) Specificity

(c) Predictive value (d) Validity

Ans: (a)

- Sensitivity represents number of true positives.

16. Screening of cervical cancer at PHC level is done by: **[DNB June 2011]**

(a) History and clinical examination (b) Colposcopy

(c) CT scan (d) PAP smear

Ans: (d)

Screening Test(s)	Disease Screened
Papanicolaou (Pap) smear test	Cervical cancer
Breast self examination (BSE)	Breast cancer
Mammography	Breast cancer

Contd.

Contd.

Bimanual oral examination	Oral cancer
ELISA, RAPID, SIMPLE	HIV (National AIDS Control Programme)
Urine for Sugar, Random blood sugar	Diabetes mellitus
AFP (alpha feto-protein)	Developmental anomalies in fetus
Digital rectal examination (DRE)	Prostate cancer
Prostate specific antigen (PSA)	Prostate cancer
Fecal occult blood test	Colorectal cancer

17. Blood screening is not done for: **[NEET Pattern 2012]**

(a) HIV (b) HBV

(c) EBV (d) HCV

Ans: (c)

- Blood sample is usually tested for HIV, HBV and HCV.

18. Screening test has the following features *except*: **[DNB Pattern 2009]**

(a) Done on apparently healthy individuals (b) It is less accurate

(c) Test results are arbitrary and final (d) It can be used as a basis for treatment

Ans: (d)

- Early diagnosis can be done using screening rather it is not used as basis for treatment.

19. For the calculation of positive predictive value of a screening test, the denominator is comprised: **[AIPGME 99, 03]**

(a) True positives + False negatives (b) False positives + True negatives

(c) True positives + False positives (d) True positives + True negatives

Ans: (c)

- PPV is defined as no of true positives out of all positives.
- So denominator is True positives + False positives.

$$PPV = \frac{a}{a+b} \times 100 = \frac{TP}{TP+FP} \times 100$$

PPV α Prevalence; Specificity; Sensitivity

20. All comprise inherent properties of a screening test *except*: **[AIIMS 2008]**

(a) Sensitivity (b) Specificity

(c) Yield (d) Predictive accuracy

Ans: (c)

- Sensitivity, Specificity and Predictive accuracy are inherent properties of screening.

21. Reliability of a test means: **[AIIMS 2016]**

(a) It measures what it is supposed to measure

(b) It is able to correctly predict the presence of disease

(c) It is able to correctly exclude the possibility of disease

(d) It yields same reading when repeated under same conditions

Ans: (d)

REPEATABILITY (RELIABILITY. PRECISION. REPRODUCIBILITY):

- Test must give consistent results when repeated more than once on same or material, under same conditions.

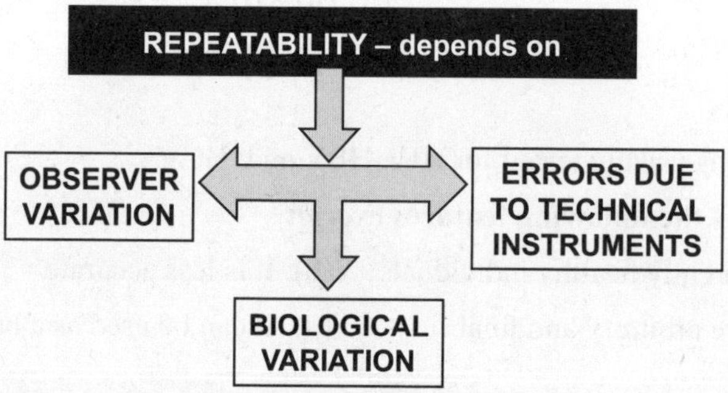

22. Reliability of a screening test does not means: **[AIIMS 2014]**

(a) Reproducibility

(b) Precision

(c) Repeatability

(d) Validity

Ans: (d)

- Refer above question.

23. True about the following is: **[NEET Pattern 2014]**

	MI present	MI absent
Positive ECG	300	100
Negative ECG	25	75

(a) Sensitivity is less than specificity

(b) Sensitivity is same as specificity

(c) PPV is more than NPV

(d) PPV is same as NPV

Ans: (d)

Sensitivity = (300/300 +25) *100= 92%

Specificity= (75/75+100) *100= 43%

PPV= (300/300+ 100) *100= 75%

NPV= (75/75 +25) *100= 75%

- Here after applying formulas, PPV = NPV.
- Sensitivity is more than specificity.

24. Positive predictive value is most affected by: **[NEET Pattern 2009]**

(a) Prevalence (b) Sensitivity

(c) Specificity (d) Relative risk

Ans: (a)

- *If sensitivity and specificity of a test are constant, predictive values are largely dependent on **prevalence of disease** in the population in which diagnostic test is being applied.*

25. A diagnostic test for a particular disease has a sensitivity of 0.90 and a specificity of 0.90. A single test is applied to each subject in the population in which diseases population is 10%. What is the probability that a person positive to this test, has disease? **[AIIMS 2005]**

(a) 90% (b) 81%

(c) 50% (d) 91%

Ans: (c)

Sensitivity = 0.90 = 90%

Specificity = 0.90 = 90%

Prevalence = 10%

$$PPV = \frac{[\text{Sensitivity} \times \text{Prevalence}]}{[\text{Sensitivity} \times \text{Prevalence}] + (1 - \text{Specificity})\,(1 - \text{Prevalence})} \times 100$$

Applying this,

- [90*10] – Numerator
- (90*10) + (100 – 90*100 – 10) – Denominator
- PPV = 50%

26. In a group of patients presenting to a hospital emergency with abdominal pain, 30% of patients have acute appendicitis. 70% of patients with appendicitis have a temperature greater than 37.5 degree Celsius and 40% of patients without appendicitis have a temperature greater than 37.5 degree Celsius. Considering these findings, which of the following statements is correct? **[AIIMS 2014]**

(a) Sensitivity of temperature > 37.5 degree Celsius as marker for appendicitis is 21/49

(b) Specificity of temperature > 37.5 degree Celsius as marker for appendicitis is 42/70

(c) PPV of temperature > 37.5 degree Celsius as marker for appendicitis is 21/30

(d) Specificity of the test will depend upon prevalence of appendicitis in applied population

Ans: (b)

- **Prevalence = 30%**
- All positive disease include 30% **(D+)**
- If 100% is population assumed then rest without disease is 70% **(D-)**
- True positive of diseased = 70% of D+ = 70 × (30/100) = 21

- False positive percentage = 40% of D- = 40 × (70/100) = 28
- False positive = 28
- **Construct table with this**

	D+ (30%)	D- (70%)
Test+	TP (21)	FP (28)
Test–	Rest (9)–FN	Rest (42)–TN

- Specificity = (TN/TN+ FP) × 100 = (42/70) × 100
- Sensitivity = (TP/TP+FN) × 100 = (21/30) × 100

27. Specificity of a screening test is the ability of a test to detect: **[AIIMS 2006, NEET Pattern 2010]**

(a) True positives (b) False positives
(c) False negatives (d) True negatives

Ans: (d)

- Specificity of a screening test is the ability of a test to detect True negatives.

28. Diagnostic power of the test is reflected by: **[NEET Pattern 2015]**

(a) Sensitivity (b) Specificity
(c) Predictive value (d) Population attributable risk

Ans: (c)

- Diagnostic power of a screening test to correctly identify a disease: Positive predictive value.
- Diagnostic power of a screening test to correctly exclude a disease: Negative predictive value.

29. If the Hemoccult test is negative for screening of colonic cancer, no further test is done. If hemoccult test is positive the individual will have a second stool sample tested with Hemoccult II test. If this second for blood, the individual will be referred for more extensive evaluation. Effect of net sensitivity and net specificity of this method of screening is: **[AIIMS 2005]**

(a) Net sensitivity and net specificity are both increased
(b) Net sensitivity is decreased and net specificity is increased
(c) Net sensitivity is increased and net specificity is decreased
(d) Net sensitivity remains the same and net specificity is increased

Ans: (b)

Net Remarks	Tests in Series (Sequential)	Tests in Parallel (Simultaneous)
Net sensitivity	Decreases	Increases
Net Specificity	Increases	Decreases
Net PPV	Increases	Decreases
Net NPV	Decreases	Increases

30. A screening test is used in same way in two similar populations, but the proportion of false positive results among those who test positive in population A is lower than among those who test positive in population B. What is the likely explanation for this finding?
[AIIMS Nov 2008, AIIMS May 03, AIPGME 01]

(a) The specificity of the test is lower in population A

(b) The prevalence of disease is lower in population A

(c) The prevalence of disease is higher in population A

(d) The specificity of the test is higher in population A

Ans: (c)

- False positive results of those who test positive in population A is lower than those who test positive in population B.
- Since false +ve is lower, then prevalence of disease will be higher in this population A.
- Test which has very high sensitivity will have high FP.

31. A test for hepatitis C is performed for 200 patients with biopsy proven disease and 200 patients known to be free of the disease. Test shows positive results on 180 of the patients with the disease, and negative results on 150 of the patients without the disease. Among those tested, this test therefore: **[NEET Pattern 2012]**

(a) Has a positive predictive value of 90%

(b) Has a negative predictive value of 75%

(c) Has a sensitivity of 90%

(d) Has a specificity of 82.5%

Ans: (c)

	D+	D–
Test+	180 (TP)	50 (FN)
Test –	20 (FN)	150 (TN)

- **Sensitivity** = (TP/TP+FN) × 100 = (180/200) × 100 = **90%**

32. Due to an effective prevention program, prevalence of an infectious disease in a community has been reduced by 90%. Physician continues to use same diagnostic test for disease that she has always used. How have test's characteristics changed? **[NEET Pattern 2012]**

(a) Its sensitivity has increased

(b) Its positive predictive value has increased

(c) Its negative predictive value has increased

(d) The test's characteristics have not changed

Ans: (c)

- **PPV α Prevalence of disease.**
- **NPV α 1/ Prevalence of disease.**

33. Validity of a test is based upon all *except*: **[AIIMS 2015]**

(a) Sensitivity

(b) Specificity

(c) Precision

(d) Accuracy

Ans: (c)

INHERENT PROPERTIES OF A SCREENING TEST

- Sensitivity
- Specificity
- Predictive accuracy

34. Probability of a test detecting a truly positive person from population of diseased is: **[AIIMS 2008]**

(a) Sensitivity of the test
(b) Specificity of the test
(c) Positive predictive value of the test
(d) Likelihood ratio

Ans: (a)

- Sensitivity of a screening test detects: true positives among all diseased.

35. For the diagnosis of Deep Vein Thrombosis, 2 tests are done together, namely Impedence Plethysmography and leg scanning after injecting 1251 fibrinogen. This process will lead to: **[AIIMS 2006]**

(a) Increasing the positive predictive value
(b) Increasing the negative predictive value
(c) Increasing the pretest odds
(d) Increasing the specificity

Ans: (b)

- Two tests are done in parallel.
- So there is increase in NPV.

Net Remarks	Tests in Series (Sequential)	Tests in Parallel (Simultaneous)
Net Sensitivity	Decreases	Increases
Net Specificity	Increases	Decreases
Net PPV	Increases	Decreases
Net NPV	Decreases	Increases

36. Denominator of positive predictive value: **[NEET Pattern 2014]**

(a) Number of true negatives + number of false negatives
(b) Number of true positives + number of true negatives
(c) Number of true positives + number of false positives
(d) Number of true positives + number of false negatives

Ans: (c)

37. Which one of the following relationships shown between different parameters of a performance of a test is correct? **[AIIMS 2014]**

(a) Sensitivity = 1 – specificity
(b) Positive predictive value = 1 – negative predictive value
(c) Sensitivity is inversely proportional to specificity
(d) Sensitivity = 1– positive predictive value

Ans: (c)

- If Sensitivity increases if TP increases and FN decreases.
- If Specificity increases if TN increases and FP decreases.
- ***Sensitivity and specificity are inversely related.***

38. Usefulness of a 'screening test' in a community depends on its: [NEET Pattern 2014]

(a) Sensitivity (b) Specificity

(c) Reliability (d) Predictive value

Ans: (a)

39. Sensitivity indicates: [NEET Pattern 2017]

(a) Positivity in disease

(b) Detection of positivity cases not in disease

(c) It identify correctly those who have not in disease

(d) Negativity in disease

Ans: (a)

40. Method of choice of tuberculosis detection mass screening is: [NEET Pattern 2013]

(a) Tuberculin test

(b) Mass Miniature Radiography (MMR)

(c) Sputum smear examination by direct microscopy

(d) Sputum culture

Ans: (c)

41. Specificity of a test refers to its ability to detect: [AIIMS 2013]

(a) True positive (b) True negative

(c) False positive (d) False negative

Ans: (b)

42. In a population of 10000 people, the prevalence of a disease is 20%. The sensitivity of a screening test is 95% and specificity is 80%. The positive predictive value of test will be: [AIIMS 2009]

(a) 54.3% (b) 45.7%

(c) 15.3% (d) 98.5%

Ans: (a)

$$\text{PPV} = \frac{[\text{Sensitivity} \times \text{Prevalence}]}{[\text{Sensitivity} \times \text{Prevalence}] + (1 - \text{Specificity})\,(1 - \text{Prevalence})} \times 100$$

$$PPV = \frac{[95 \times 20]}{[93 \times 20] + [(100 - 80)(100 - 20)]} \times 100 = 54.3\%$$

43. True about reliability of a test: **[AIIMS 2015]**

(a) Gives same results on repeated tests
(b) Investigator's knowledge is important
(c) Consistency and reproducibility of the test are not a problem
(d) Extent of variation of measurement of contained behavior

Ans: (a)

REPEATABILITY (RELIABILITY, PRECISION, REPRODUCIBILITY):

- Test must give consistent results when repeated more than once on same or material, under same conditions.

44. A test has high false positive rate in a community. True is: **[AIIMS 2010]**

(a) High prevalence
(b) Low prevalence
(c) High sensitivity
(d) High specificity

Ans: (c)

False positive rate (α) = 1 – specificity

- **A screening test with a high specificity will have few false positives.**
- A test which has very high sensitivity will have high FP.

45. True regarding specificity: **[NEET Pattern 2011]**

(a) Identifies false +ve
(b) Identifies false –ve
(c) Identifies true +ve
(d) Identifies true –ve

Ans: (d)

46. A doctor order 6 tests for SLE. Which is needed for inference? **[AIIMS 2010]**

(a) Prior probability of SLE, sensitivity and specificity of test
(b) Incidence of SLE and predictivity of each test
(c) Incidence and prevalence of SLE
(d) Relative risk of SLE in the patient

Ans: (a)

- Post-test probability of a disease in a population is same as PPV.
- Pre-test probability of a disease in a population is same as Prevalence.

47. Epidemiological survey of 'at risk' is called: **[NEET Pattern 2013]**

(a) Survey
(b) Screening
(c) Surveillance
(d) Rehabilitation

Ans: (b)

48. You have clinically diagnosed a patient as having SLE and ordered 6 tests, out of which 4 have come positive and 2 have come negative. To know the probability of SLE at this point, you need to know: [AIIMS November 2011, AIPGME 2012]

(a) Incidence of SLE and predictive value of each test

(b) Incidence and prevalence of SLE

(c) Relative risk of SLE in this patient

(d) Prior probability of each test, Sensitivity and specificity of each test

Ans: (d)

- Post-test probability of a disease in a population is same as PPV.
- Pre-test probability of a disease in a population is same as prevalence.

49. A new method of measuring Haemoglobin levels has been developed. Ten successive readings of a single sample are as follows: 9.4, 10.4, 9.6, 9.1, 10.8,12.1, 10.1, 9.8, 9.2, 9.5. But Haemoglobin measured by standard calorimetry was 10.2. Therefore given method has: [AIIMS 2011]

(a) Low validity, low reliability
(b) Low validity, high reliability

(c) High validity, low reliability
(d) High validity, high reliability

Ans: (c)

- Reliability is precision (repeatability) and Validity is accuracy (close to true/actual value).

50. If prevalence of a disease in a population increases, predictive value of a positive test: [NEET Pattern 2013]

(a) Increases
(b) Decreases

(c) Remains constant
(d) Becomes compromised

Ans: (a)

PPV α Prevalence; Specificity; Sensitivity

51. Best time to screen in Breast self examination technique is: [Recent Question 2014]

(a) 1 week before the menstruation
(b) 1 week after the menstruation

(c) During ovulation
(d) 2-3 days post-ovulation

Ans: (b)

52. Screening test for Breast and Genital tract malignancy is: [AIIMS November 2014]

(a) CA-125
(b) Mammography

(c) Office endometrial aspiration
(d) Pap smear

Ans: (a)

53. Formula A/A+B in the following table denotes: **[DNB Pattern 2008]**

Test result	Persons with disease	Persons without disease	Total
Positive	A	B	A + B
Negative	C	D	C + D

(a) Specificity (b) Sensitivity
(c) PPV (d) NPV

Ans: (c)

- PPV is true positives out of all positives.

54. Sensitivity is: **[DNB Pattern 2007]**

(a) True positive/True positive + false negative
(b) True positive/False positive + true negative
(c) True negative/False negative + false positive
(d) True negative/False negative + true positive

Ans: (a)

55. Most number of false positives by a screening test is because of: **[DNB Pattern 2011]**

(a) High specificity (b) High sensitivity
(c) High prevalence (d) Low prevalence

Ans: (b)

- Test which has very high sensitivity will have high FP.

56. Ability of test to correctly diagnose % of sick people who are having condition is called as: **[DNB Pattern 2009]**

(a) Sensitivity (b) Specificity
(c) Positive predictive value (d) Negative predictive value

Ans: (a)

57. Most important factor for a test to be a good screening test is: **[DNB Pattern 2011]**

(a) Specificity (b) Sensitivity
(c) Reliability (d) Predictive value

Ans: (b)

58. Validity includes: **[NEET Pattern 2013]**

(a) Sensitivity and specificity (b) Precision
(c) Acceptibility (d) None

Ans: (a)

59. Positive predictive value is a function of sensitivity, specificity and: **[AIIMS 2014]**

(a) Incidence (b) Prevalence
(c) Negative predictive value (d) Accuracy

Ans: (b)

60. True statement about PPV is: **[NEET Pattern 2014]**

(a) It increases with prevalence
(b) It decreases with prevalence
(c) No relation with prevalence
(d) Doubles with decrease in prevalence

Ans: (a)

61. 5000 persons underwent screening for a disease. Out of 500 diseased, 350 reported True positive and out of 4500 healthy, 3000 reported True negative. Which of the following is correct about this screening test? **[Recent Question 2014]**

(a) Sensitivity 70%
(b) Specificity 70%
(c) Sensitivity 80%
(d) Specificity 80%

Ans: (a)

	D+	D–
Test+	350	1500
Test–	150	3000

- **Sensitivity = (TP/TP + FN) × 100** = (350/350+150) × 100 = 70%.

62. Sensitivity is: **[DNB Pattern 2004]**

(a) True positive/True positive + false negative
(b) True positive/False positive + true negative
(c) True negative/True negative + false positive
(d) True negative/False negative + true positive

Ans: (a)

63. A diagnostic test for a particular disease has a sensitivity of 0.90 and a specificity of 0.80. A single test is applied to each subject in the population in which diseased population is 30%. What is the probability that a person, negative to this test, has no disease? **[NEET Pattern 2014]**

(a) < 50%
(b) 70
(c) 95%
(d) 72%

Ans: (c)

$$\text{NPV} = \frac{[\text{Specificity} \times (1 - \text{Prevalence})]}{[\text{Specificity} \times (1 - \text{Prevalence})] + [(1 - \text{Sensitivity}) \times \text{Prevalence}]} \times 100$$

NPV = 80 × (100 – 30)/[80 × (100 – 30) + (100 – 90) × 30] = 95%

64. Sensitivity of a test is: **[DNB Pattern 2007]**

(a) True positive/True positive + false negative
(b) True positive/False positive + true negative
(c) True negative/True negative + false positive
(d) True negative/False negative + true positive

Ans: (a)

65. Usefulness is: **[DNB Pattern 2006]**

(a) True positive/True positive + false negative

(b) True positive/False positive + true negative

(c) True negative/True negative + false positive

(d) True negative/False negative + true positive

Ans: (a)

66. Specificity measures: **[DNB Pattern 2010]**

(a) True positive (b) True negative

(c) False positive (d) False negative

Ans: (b)

67. False negative means: **[Bihar 2006]**

(a) Persons have disease but show negative test result

(b) Persons have not disease but show negative test result

(c) Persons have disease but show positive test result

(d) Persons have not disease but show positive test result

Ans: (a)

68. A screening test is more sensitive: **[DNB Pattern 2015]**

(a) Few false positive (b) Few false negative

(c) More false positive (d) More false negative

Ans: (b)

69. Specificity means: **[DNB Pattern 2005]**

(a) True positive (b) True negative

(c) False positive (d) False negative

Ans: (b)

70. High false positives in a test is due to: **[DNB Pattern 2008]**

(a) High incidence (b) High prevalence

(c) High sensitivity (d) High specificity

Ans: (c)

71. High false positive cases in a community signify that disease has: **[DNB Pattern 2008]**

(a) High prevalence and Low incidence (b) High incidence and Low prevalence

(c) Low prevalence and Low incidence (d) High incidence and High prevalence

Ans: (b)

72. Sensitivity numerator is: **[DNB Pattern 2011]**

(a) False positives (b) False negatives

(c) True negatives (d) True positives

Ans: (d)

73. Positive predictive value of a test does not depend upon: **[DNB Pattern 2011]**

(a) Sensitivity (b) Specificity

(c) Prevalence of disease (d) Incidence of disease

Ans: (d)

74. A screening test was positive in 50% of diseased and 10% of healthy population. What is the specificity of the test? **[MP 2006]**

(a) 0.5 (b) 0.9

(c) 0.83 (d) 0.064

Ans: (b)

	D+	D–
Test+	50	10
Test–	50	90

- **Specificity = (TN/FP+TN) × 100** = (90/100) × 100 = 0.9

75. True positive cases are detected by: **[DNB Pattern 2010]**

(a) Specificity (b) Sensitivity

(c) Positive predictive value (d) Negative predictive value

Ans: (b)

76. Specificity means: **[DNB Pattern 2012]**

(a) True positive (b) True negative

(c) False positive (d) False negative

Ans: (b)

77. Most reliable test for screening of diabetes mellitus: **[NEET Pattern 2012]**

(a) GTT (b) Glycosylated hemoglobin

(c) Fasting blood sugar (d) Urine for sugar

Ans: (c)

78. Most specific screening test for Vitamin D deficiency is: **[NEET Pattern 2013]**

(a) 7-dehydrocholesterol (b) 1, 25 dihydroxy Vitamin D

(c) 25 hydroxy Vitamin D (d) Serum calcium levels

Ans: (c)

PART 3: EPIDEMIOLOGY

1. BASICS OF EPIDEMIOLOGY

DEFINITION:

- Defined as is study of distribution and determinants of health related states or events in specified population and its application to control of health problems.
- It involves study of Disease frequency, Distribution and Determinants.

2. TYPES OF EPIDEMIOLOGICAL STUDIES

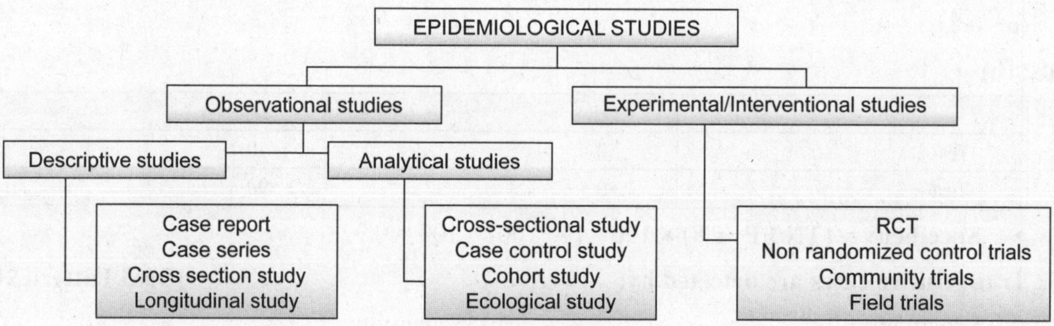

FLOWCHART

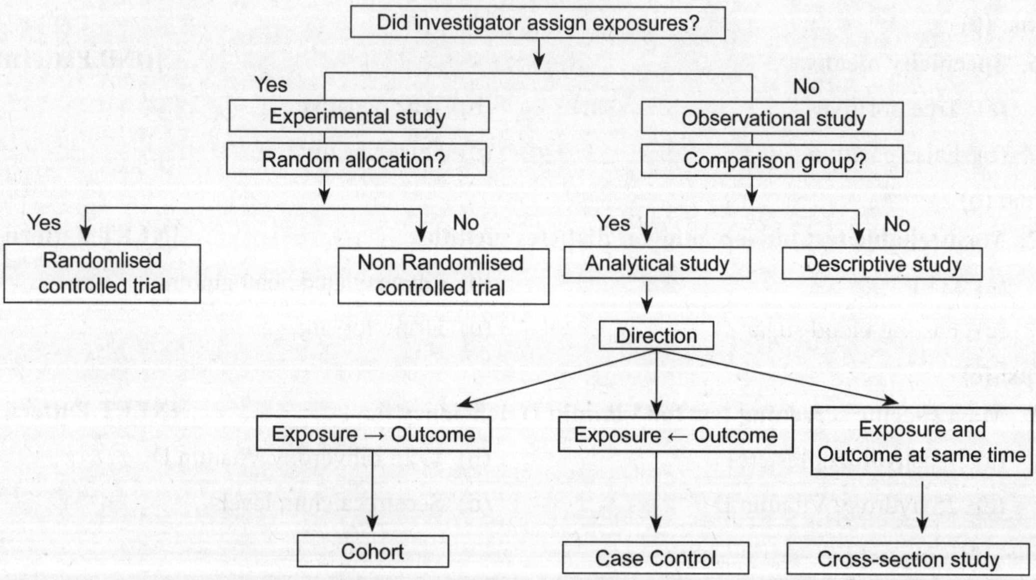

- Descriptive studies – Generates an Hypothesis
- Analytical studies – Tests the Hypothesis
- Experimental studies – Provides Solution

3. DESCRIPTIVE STUDIES

EPIDEMIC:

- It is occurrence in a community or region of cases of an illness/health-related events clearly *in excess of normal expectancy.*
- Epidemic threshold = Endemic frequency + 2 Standard Error.

TYPES OF EPIDEMICS

- **Common source epidemics**
  - Single exposure or "point source" epidemic
  - Continuous or multiple exposure epidemic
- **Propagated epidemics**
  - Person-to-person
  - Arthropod vector
  - Animal reservoir
- **Slow (modern) epidemics**

Epidemic curve: *It is a graphical representation of the time distribution of epidemic cases.*

SEASONAL TREND:

- Related to environmental conditions (e.g., Temperature, Humidity, Rainfall, Overcrowding etc.) that directly or indirectly favour disease transmission.
- **Example:** Gastrointestinal infections are prominent in summer months.

CYCLIC TREND:

- It is due to naturally occurring variations in herd immunity.
- A build-up of susceptible is required in "herd" before there can be another attack.
- Diseases occur in cycles spread over short periods of time (days, weeks, months or years).
- **Example:** Measles in prevaccination era occurred in cycles with major peaks every 2-3 years and rubella every 6-9 years.

LONG-TERM OR SECULAR TREND:

- It is change in occurrence of disease (i.e., progressive increase or decrease) over a long period of time, generally several years or decades.
- Consistent tendency to change in particular direction or definite movement in one direction.
- **Example:** Coronary heart disease, lung cancer and diabetes (consistent upward trend in developed countries).

OBSERVATIONAL STUDIES

1. **CASE REPORT:**
   - Report of a **single patient** with the rare diagnosis or unexpected course of the event.
2. **CASE SERIES:**
   - Report of a group of patients with the same diagnosis.

CASE SERIES AND CASE REPORTS

Advantages:

- Can aggregate cases from disparate sources to generate hypotheses and describe new syndrome.
- **Example:** Hepatitis, AIDS.

Disadvantages:

- Cannot test for statistical association as there is no relevant comparison group.
- Based on individual exposure

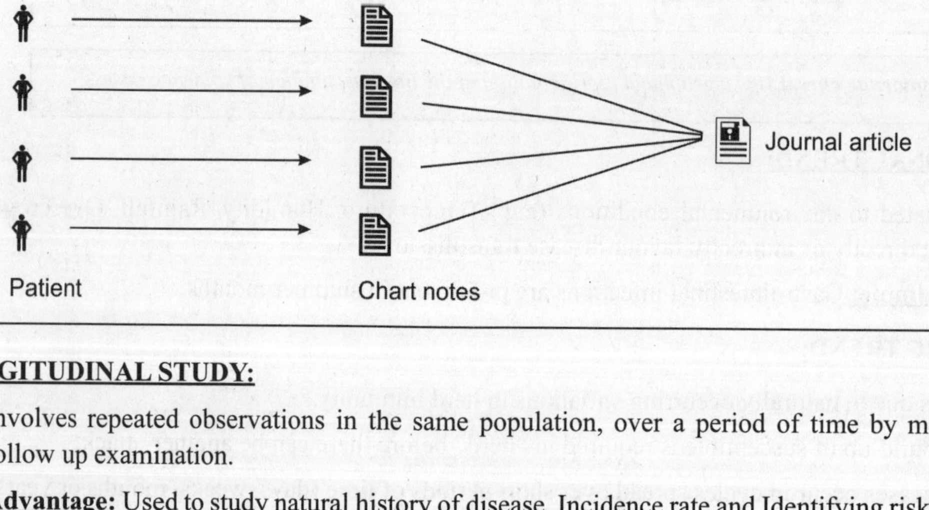

3. **LONGITUDINAL STUDY:**
   - Involves repeated observations in the same population, over a period of time by means of follow up examination.
   - **Advantage:** Used to study natural history of disease, Incidence rate and Identifying risk factors of disease.
   - **Disadvantage:** Difficult to organize and Time consuming.

4. ANALYTICAL STUDIES

1. COHORT STUDY (FOLLOW UP STUDY):

- It is a type of analytical (observational) study used for 'hypothesis testing'.
- It is a 'Prospective' study design.
- Proceeds from Cause to effect (or) Exposure to outcome.

Design of a Cohort Study

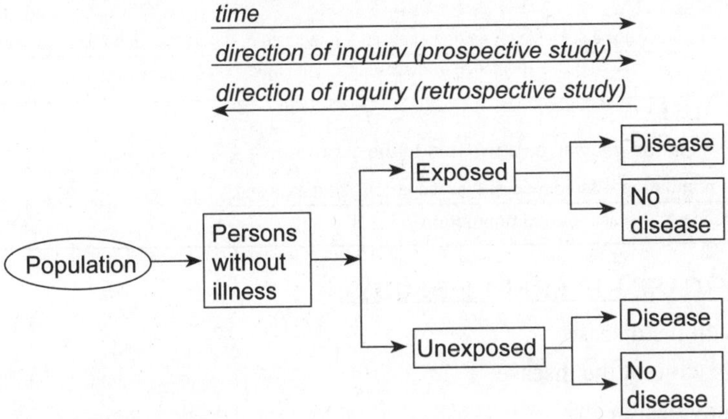

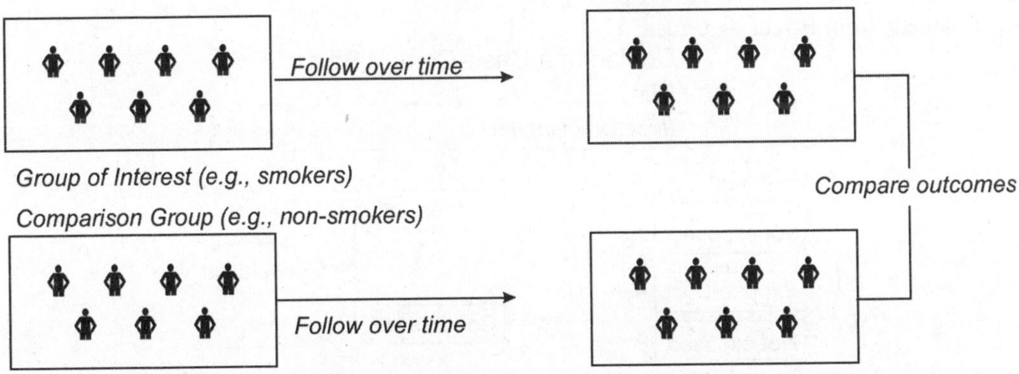

APPLICATION OF COHORT STUDY

RELATIVE RISK (RISK RATIO)

- Ratio of the incidence rates between exposed group and non-exposed group.
- Can only be calculated from prospective studies (cohort studies).
- **Ratio that measures strength of association**

$$RR = \frac{\text{Incidence rate among exposed}}{\text{Incidence rate among non-exposed}}$$

Contd.

Contd.

INTERPRETATION OF RR:

RR > 1 = Increased risk of outcome among "exposed" group
RR < 1 = Decreased risk, or protective effects, among "exposed" group
RR = 1 = No association between exposure and outcome

ATTRIBUTABLE RISK (AR):

- Indicates to what extent the disease under study can be attributed to the risk/exposure.
- It is additional incidence of disease that can be attributed to the risk factor in question.
- Amount of disease that might be eliminated if factor under study is controlled/eliminated.
- Usually expressed as a percentage

$$\text{Attributable risk [Risk difference]} = \frac{\text{[Incidence of disease/death among exposed} - \text{Incidence of disease/death among non exposed]}}{\text{Incidence of disease or death among exposed}}$$

POPULATION ATTRIBUTABLE RISK:

- It indicates the risk in the total population that can be attributed to the exposure.

$$\frac{\text{Incidence of disease in total population} - \text{Incidence of disease among non exposed}}{\text{Incidence of disease in total population}}$$

2. CASE CONTROL STUDY (CASE REFERENCE STUDY):

- **Cases:** Study subjects with the disease
- **Controls:** Study subjects without the disease
- It is a 'Retrospective' study design
- Proceeds from Effect → Cause

Design of a Case-Control Study

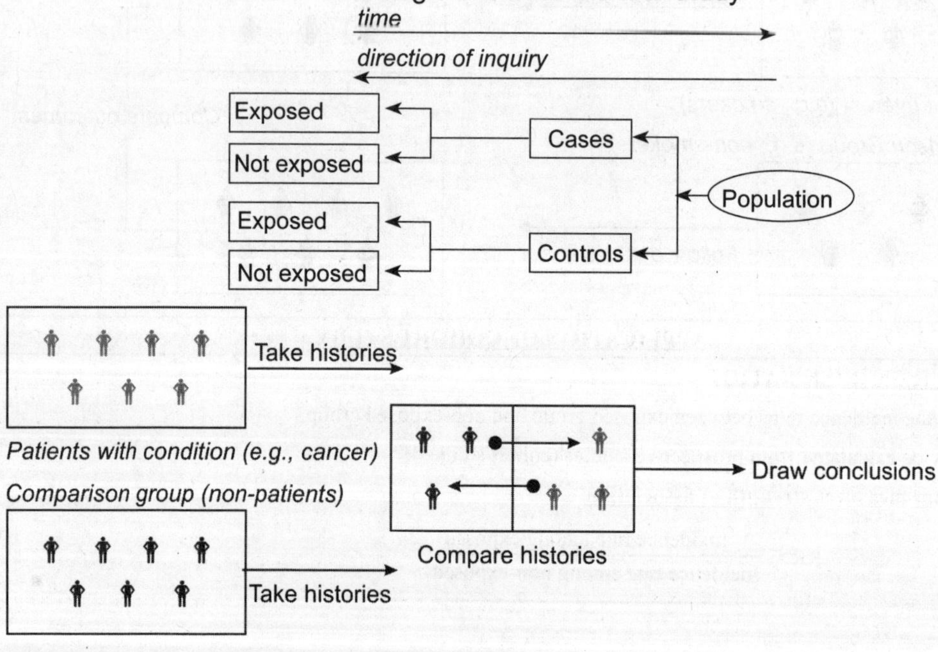

NESTED CASE-CONTROL STUDIES
• Nested case-control design makes cohort studies less expensive. • Cases and controls are both chosen from a defined cohort, for which some information on exposures and risk factor is already available. • Additional information on new cases and controls, selected for study, is collected and analysed. • Useful when measurement of exposure is expensive.

APPLICATION OF CASE CONTROL STUDY

- **Odds** is probability of an event occurring vs probability of it not occurring, in other words
- OR is a measure of the strength of association between a risk factor and an outcome.

$$\text{Odds Ratio} = \frac{\text{Odds that case was exposed}}{\text{Odds that control was exposed}}$$

$$\text{Odds ratio} = ad/bc$$

THREE ASSUMPTIONS TO CALCULATE OR
• Disease must be relatively rare. • Cases must be representative of the diseased. • Controls must be representative of the non-diseased.

- Incidence of the disease in exposed and unexposed are not available, hence relative risk cannot be calculated in a case control study.
- So Odds Ratio is calculated as **'an estimate of Relative Risk'.**

Interpretation of Odds Ratio

OR value	Association	Interpretation
OR < 1	Negative association	**Negative association** exists between the disease and exposure (exposure is protective)
OR = 1	No association	**No association** exists between the disease and exposure (no role of exposure on outcome)
OR > 1	Positive Association	**Positive association** exists between the disease and exposure (exposure in harmful)

3. CROSS-SECTIONAL STUDY (PREVALENCE STUDY):

- Simplest form of an observational study (Descriptive / Analytical).
- Based on a **single examination** of a cross-section of population at one point of time.
- Provides very little information about natural history of disease or about the rate of occurrence of new cases (incidence).
- Measures **disease and exposure simultaneously** in a well-defined population.
- **Odd's Ratio (OR) can be calculated in an analytical cross-sectional study.**

Advantages:

- Good for identifying **prevalence** of common diseases like Diabetes, HTN.

Disadvantages:

- Cannot determine whether exposure preceded disease.
- It considers prevalent rather than incident cases, results will be influenced by survival factor.

4. ECOLOGICAL STUDY (CO-RELATIONAL STUDY):

- Uses measures that represent characteristics of entire populations.
- It describes outcomes in relation to age, time, utilization of services, or exposures.

Advantages:

- Generates hypotheses for case-control studies and environmental studies

Disadvantages:

- Data are obtained from group, disease and exposure in individual can't be linked
- Potential confounders cannot be controlled
- Dose-response relationship cannot be determined
- Caution must be taken to avoid drawing inappropriate conclusions/ecological fallacy

	Case control study	Cohort study
1	Proceeds from ***"effect to cause"***. Exposure and outcome has occurred before start of investigation	Proceeds from ***"cause to effect"***. Outcome has not yet occurred at the time the investigation begins. Cohorts are identified prior to appearance to disease under investigation
2	Investigator goes back in time to select his study group and traces them through time to occurrence of effect/outcome	Begins in present and continues into future. Tests whether the disease occurs more frequently in those exposed, then in those not similarly exposed
3	No Attrition	Attrition is a major problem
4	Yields quick results and relatively easy to do	Study takes a long time
5	Involves fewer numbers of subjects	Involves large number of subjects
6	Suitable for study of rare diseases	Suitable for rare exposure
7	Cheaper	Costly
8	No follow up	Follow up is an integral part
9	Yields only an estimate of RR, i.e., odds ratio	Yields incidence rate, RR and AR, PAR
10	Can yield information about several aetiological factors or risk factors	Can yield information about more than one disease outcome from single aetiological factors
11	1st approach to testing a causal hypothesis	
12	***Disadvantage***–Recall bias, Selection bias	No recall bias

MUST REMEMBER PICTURE

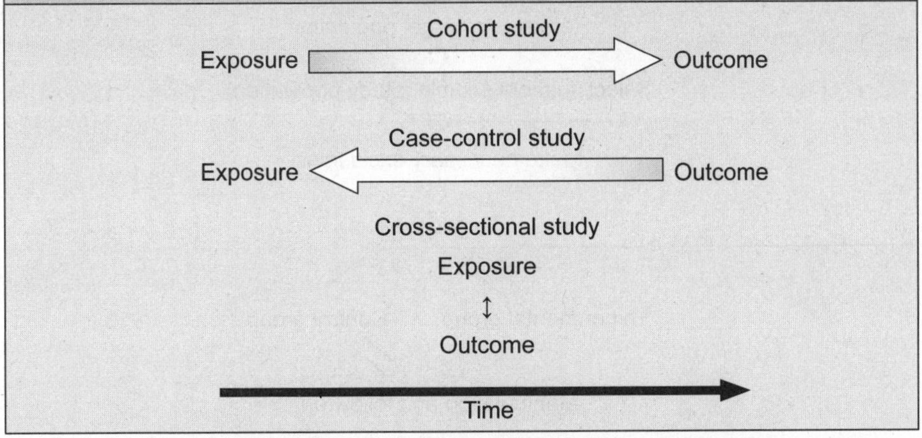

EXPERIMENTAL/INTERVENTIONAL STUDIES

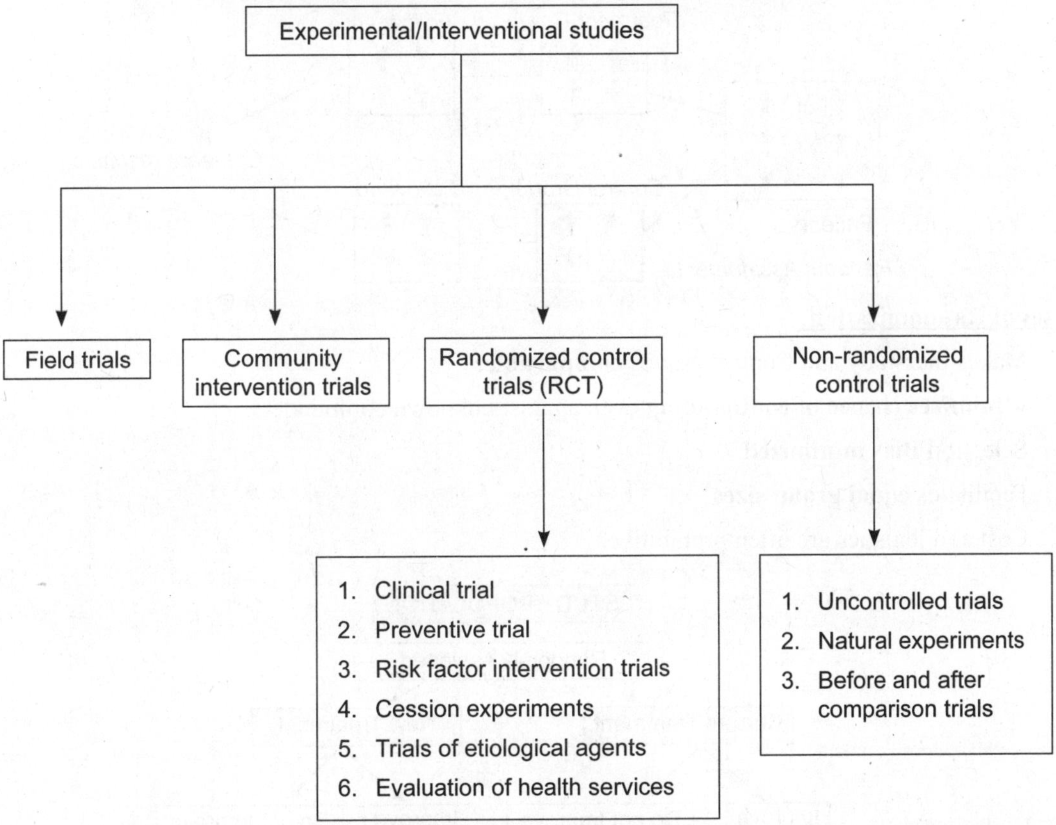

1. **RANDOMIZED CONTROLLED TRIAL:**

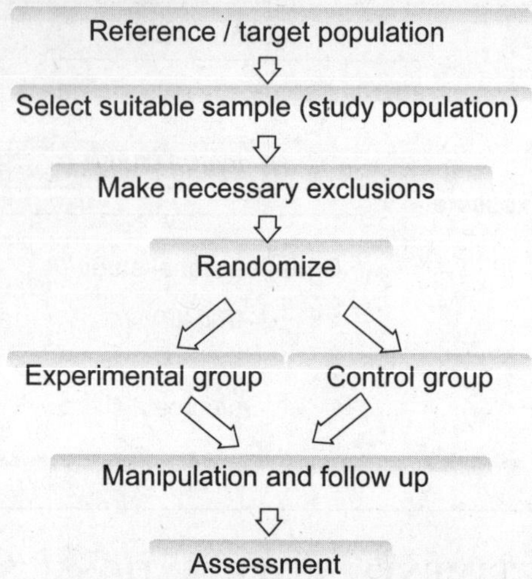

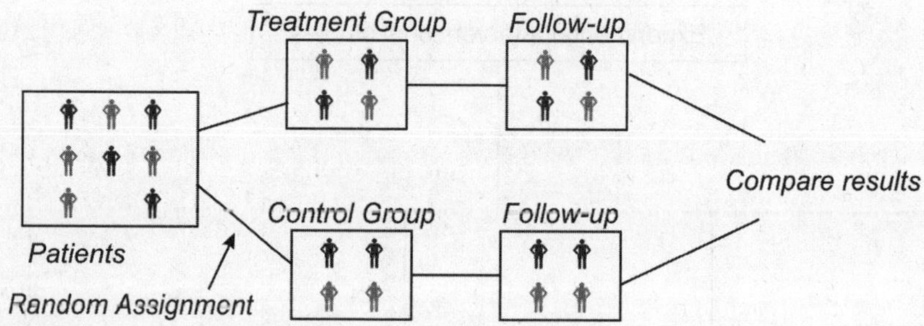

Uses of Randomization:

- Makes the Study and Control groups **comparable**.
- **Minimizes** chance of **confounding** even against unknown confounders.
- Selection **Bias minimized**.
- Facilitates **equal group sizes**.
- Cost and logistics are often prohibitive.

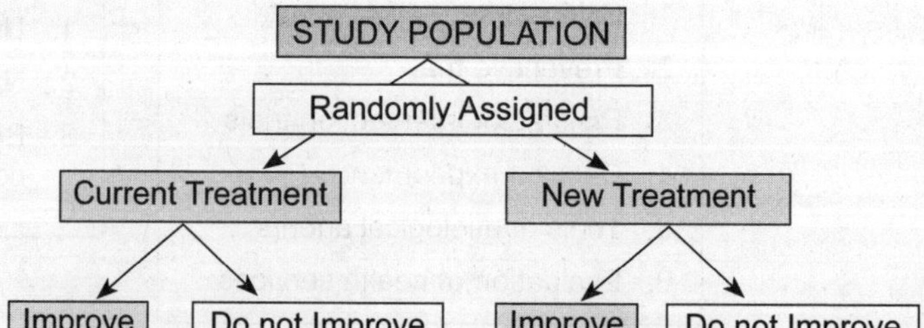

Best Method for Evaluation of New Programme or New Therapy

BLINDING

- **Single Blind:** Participant does not know which treatment received
- **Double Blind:** Observer and Participant, both do not know which treatment received
- **Triple Blind:** Person analyzing data, Observer and participant, don't know which treatment received

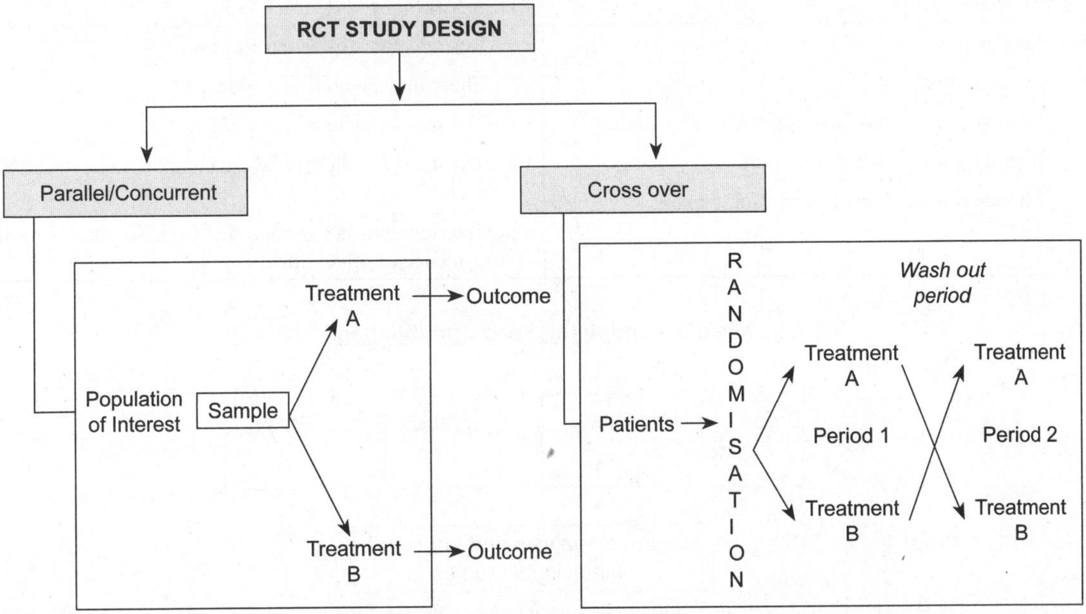

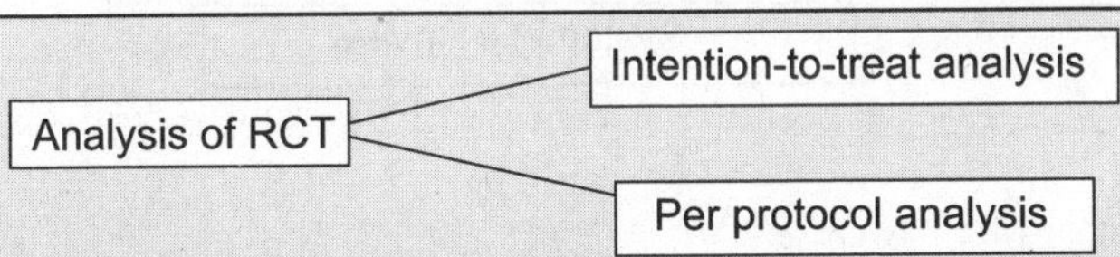

INTENTION-TO-TREAT ANALYSIS	PER PROTOCOL ANALYSIS
All subjects randomly allocated to exposure groups are analysed, irrespective of whether they completed treatment or not.	Compares the outcomes in only those subjects who complete treatment protocols in different groups. Analysis restricts to population who strictly adhered to protocol
Advantages: Preserves benefits of randomization Because good and poor compliers differ, it helps ensure that study results are unbiased Gives information on effectiveness of treatment under everyday practice conditions.	**Advantages:** Determines treatment effects under ideal conditions. **Disadvantages:** Comparison groups may not be similar in size.

2. SYSTEMIC REVIEW:

- A systematic review is a systematic assembly, critical appraisal and synthesis of all relevant studies on a specific topic.

PRISMA (Preferred Reporting Items for Systematic Reviews and Meta-Analysis)
It is an evidence based minimum set of items for reporting in systematic reviews and meta-analyses.

Advantages	Disadvantages
• Reduce bias • Replicability • Resolve controversy between conflicting studies • Identify gaps in current research • Provide reliable basis for decision making	• Results may still be inconclusive • There may be no trials/evidence • The trials may be of poor quality • The intervention may be too complex to be tested by a trial • Practice does not change just because you have the evidence of effect/effectiveness

Includes both **published** and **unpublished** studies

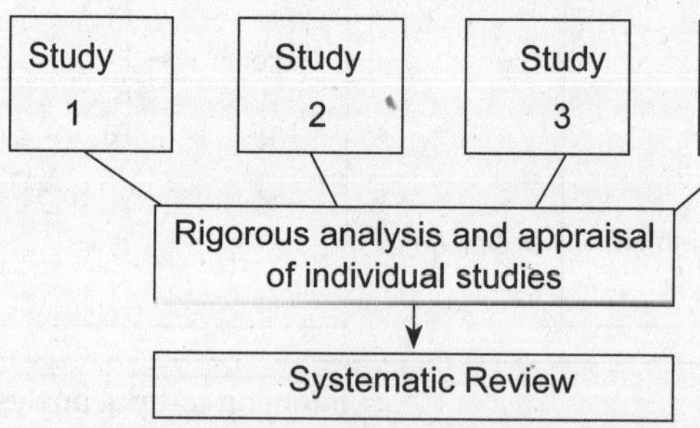

3. META-ANALYSIS:

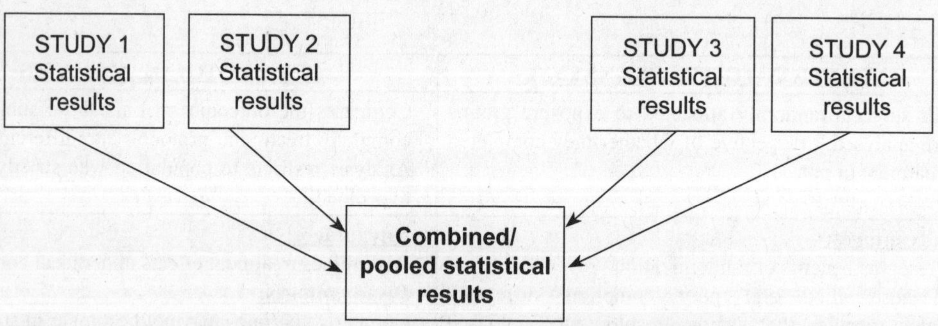

- It is statistical synthesis of data from separate but similar (comparable) studies [particularly RCT] leading to a quantifiable summary of the pooled results to identify overall trend.
- In meta-analysis no new data is collected.

- It combines results of several trials, each of which may deal with a relatively small sample, to obtain a better overall estimate of effect.
- It is a retrospective research, subject to methodological deficiencies of each included study.
- It allows comparisons to be made between studies even if they used different measures of outcome.
- Use has increased in recent years for ethical reasons, cost issues and need to have an overall idea of effects of a particular intervention in different population groups.

Relative ability to "prove" causation in different types of Epidemiological study:

- Meta-analysis > RCT > Cohort > Case control > Cross-sectional > Ecological

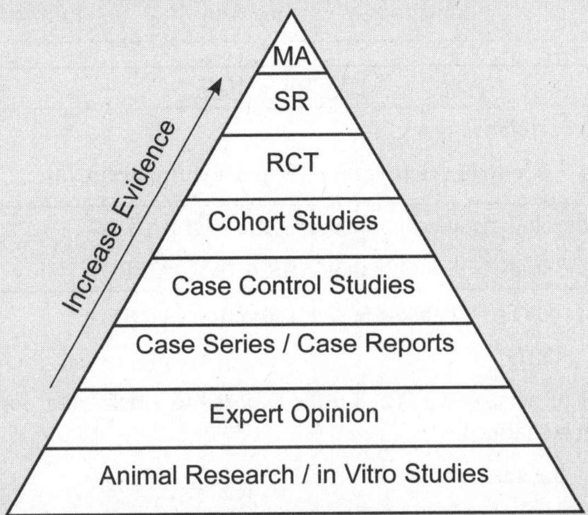

Table: Advantages and disadvantages of various study designs

Study Design	Advantages	Disadvantages
Primary data		
Randomized clinical trial	Highest level of evidence possible, least prone to bias and error	Costly, time consuming; can be ethical concerns for randomization in certain scenarios (i.e., surgery)
Cohort study	Can directly observe outcomes in association with risk factors or exposures: can directly measure risk	Costly, time consuming; patients who are lost to follow-up can be a major source of bias
Case control study	Can estimate outcomes associated with exposure/risk factors in a much less costly and time-consuming approach than a cohort study	Can be very difficult to identify the appropriate control group, which can lead to bias, error, and wrong conclusion
Cross-sectional study	Can rapidly obtain data comparing exposures/risk factors to outcomes at one point in time (prevalence)	Only gives data from one time point, cannot assess risk over time (incidence)

Contd.

Contd.

Secondary data		
Case report/Case series	Easy, inexpensive; can rapidly explore hypothesis and association between exposures/risk factors and an outcome	Very prone to selection bias, retrospective design can impact data quality, lack of a control group can make conclusions uncertain
Meta-analysis/systematic review	Can compile all available data in one place for complete analysis and thorough evaluation of weakness and limitations of existing data	Prone to public bias; garbage in, garbage out principle, complex mathematics can make conclusions difficult to interpret

EXTRA EDGE

- Relative risk reduction = 1 – relative risk
- **Absolute risk reduction = (Control event rate) – (Experimental event rate).**
  - If the outcome of the study is undesirable then ARR = CER – EER
  - If the outcome of the study is desirable then ARR = EER - CER
- **Number needed to treat (NNT) = 1/ absolute risk reduction (ARR)**

  Number needed to harm (NNH)

  a. NNH is the number of persons who need to be exposed to a risk factor for one person to be harmed who would otherwise not be harmed.

  b. NNH is 1 divided by the attributable risk.
- Hazard ratio (HR) is similar to relative risk but is used when risk is not constant to time. It is typically used when analysing survival over time.

QUASI EXPERIMENTAL STUDY:

- Experimental study design, where for reasons of practicability or Ethics, one or more of 3 essential elements of RCT [i.e., Randomization, Controls and Blinding] are not observed.
- **Example**: Non randomised "concurrent" trial, Uncontrolled trials, Before and after trial, Natural experiments.

5. BIAS AND CONFOUNDING FACTOR

Selection Bias	Measurement Bias
Self-Selection (Volunteers) Bias	Recall Bias
Berksonian Bias	Detection Bias
Survivorship (Neyman's) Bias	Observer's Bias
Healthy Worker Effect	
Exposure-Related Bias	
Loss to Follow - Up Bias	
Inappropriate Control Group	

BIAS

- **Confounding bias:** Defined as one which is associated both with exposure and disease, and is unequally in study and control groups.
- **Memory or recall bias:** When cases and controls are asked questions about their past history, it may be more likely for the cases to recall existence of certain events.
- **Selection bias:** May be difference in characteristics between cases and controls.
- **Berkesonian bias:** Different rates of admission to hospitals for people with different diseases.
- **Interviewer's bias:** Bias also may occur when interviewer knows hypothesis and also knows who the cases are.
- **Nonresponse bias:** Occurs because individuals who do not respond to a call to participate in research studies are generally different from those who do respond.
- **Self selection bias:** Self referral of the persons for the study affects the validity because the reasons for self referral may be associated with outcome under study.
- **Measurement bias:** Occurs when the individual measurements or classifications of disease or exposure are inaccurate.
- **Recall bias:** A form of measurement bias of particular importance in retrospective case-control studies. This occurs because of differential recall of information by cases and controls. Recall bias can either underestimate or exaggerate degree of effect associated with exposure.
- **Information bias:** Bias in estimating an effect can be caused by measurement errors in needed information. Such bias is called as information bias.
- **Sampling bias:** Sample is collected in such a way that some members were not equally likely to have been selected.
- **Publication bias:** Tendency for published studies to be systematically different from all completed studies of a question. In simple words (Hiding the results of a drug's adverse effects while publishing) – at the end of the RCT. Published articles are more likely to be positive because of a pervasive preference for positive results.

Measures to Reduce Biases:

- Appropriate designing of the study
- Validation of the tool/instrument (calibration)
- Training of the investigators/observers (to reduce intra observer variations)
- Usage of kappa statistic to reduce inter-observer variations
- Randomization of selection of study subjects (to reduce volunteer bias)
- Blinding: Single/double/triple

CONFOUNDING FACTOR

- **Confounder** is any factor associated with both **exposure and outcome**, and has an **independent effect** in causation of outcome (thus it is a risk factor by itself).
- Confounder is found **unequally** distributed between the study and control groups.

Measures to reduce confounding during the study:
• Randomization • Restriction • Matching

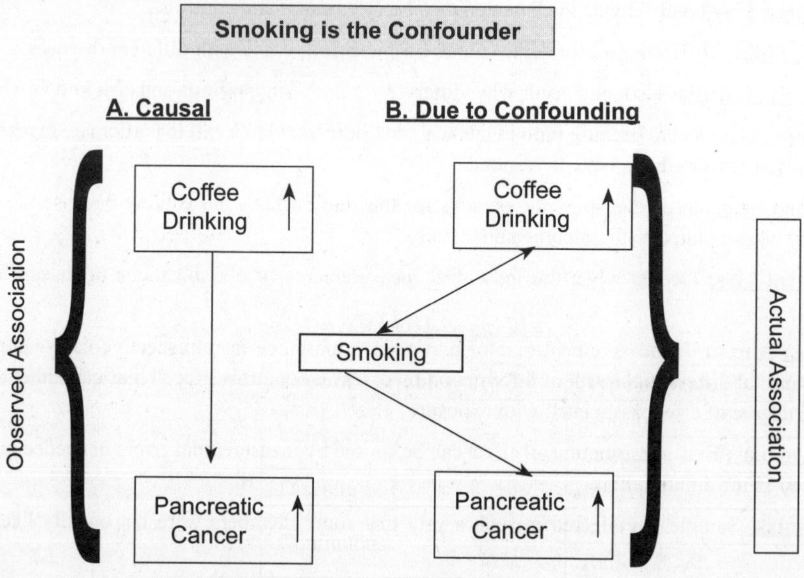

- Confounding may be considered as confusion of effects.
- Confounders are the extraneous factors that are responsible for difference in disease frequency between the exposed and unexposed.
- **Criteria for confounding factor**
  - A confounding factor must be an **extraneous risk factor** for disease.
  - A confounding factor should be **associated with** the exposure under study in source population.
  - A confounding factor must not be affected by the exposure or disease.

CONFOUNDING IS CONTROLLED BY FOLLOWING METHODS

- **Randomization** – ensuring confounding variables are equally distributed among groups being compared
- **Restriction** – limiting the study to people who have particular characteristics
- **Matching** – ensuring potential confounding variables are evenly distributed in 2 groups being compared
- **Stratification** – confounding occurring during analysis phase rather than design phase – measurement of strength of association in well defined and homogenous categories (strata)
- **Statistical modeling** – multivariate analysis – done to estimate strength of associations while controlling for number of confounding variables

PHASE TRIALS

Phase	Unit of Study	Purpose
PRECLINICAL PHASE		
Lab experiments	Animals	Pretesting
CLINICAL PHASES		
Phase 0	Healthy human volunteers	Micro-dosing
Phase I	Healthy human volunteers	Safety and non-toxicity profile
Phase II	Patients	Effectiveness
Phase III	Patients	Comparison with existing drugs
Phase IV	Patients	Long term side effects

MULTIPLE CHOICE QUESTIONS

1. Concerning anti-cancer clinical drug trials which statement is correct?

(a) Phase 1 clinical trials usually involve large numbers of patients to test for anti cancer properties

(b) Phase 2 trials will aim to ascertain the response of the cancer to treatment

(c) A Phase 3 trial testing a new cancer drug will assess its cost-effectiveness

(d) Patient consent is not required for tie inclusion of patients with advanced cancer into Phase 1 trials only

Ans: (b)

- Phase 1 is testing new treatment for assessing dose range side effects and effects involves only small number of volunteers.
- Phase 2 is usually larger than phase 1 is for showing whether new treatment is better than existing.
- **Phase 3 is comparison between treatment with new therapy compared with standard treatment–compares survival or side effects.**

Phases of Clinical Trials:

	Objective	Unit of study	Sample size
Phase 0	Pre-clinical Phase/Animal research	Lab animals	
Phase I	Safety and Acceptability	Healthy volunteers	Upto 50
Phase II	Log term safety Dose and Schedule Efficacy	Low Risk patients	100-500
Phase III	Effectiveness	High Risk patients	> 1000
Phase IV	Post Marketing Surveillance	Community based	> 1000

2. A randomized controlled trial is conducted to evaluate a novel therapy. What level of evidence does this represent? [DNB Pattern 2015]

(a) 1b

(b) 2a

(c) 2b

(d) 3

Ans: (a)

Explanation:

Levels of evidence for studies evaluating therapy, prevention, aetiology or harm:

1a	–	Systematic review of RCTs
1b	–	Individual RCT
2a	–	Systematic review of cohort studies
2b	–	Individual cohort study
3a	–	Systematic review (with homogeneity) or case control studies
3b	–	Individual case control study
4	–	Case series
5	–	Expert opinion

3. **True about confounding factor is:** **[AIIMS 2009]**
   (a) It is found equally between the study and the control groups
   (b) It is itself a risk factor for the disease
   (c) Confounding can be eliminated by selecting a small group
   (d) It is associated either with the exposure or the disease

Ans: (b)

4. **Nested case control study is a type of:** **[AIIMS 2009]**
   (a) Retrospective study (b) Prospective study
   (c) Descriptive study (d) Cross-sectional study

Ans: (b)

5. **Recall bias is most commonly associated with which study design?** **[AIIMS 2009]**
   (a) Case control study (b) Cohort study
   (c) Randomised controlled trial (d) Cross-sectional study

Ans: (a)

6. **Incidence can be calculated in:** **[AIIMS 2010]**
   (a) Case-control study (b) Prospective study
   (c) Retrospective study (d) Cross-sectional study

Ans: (b)

7. **Confounding bias is reduced by all *except*:** **[AIIMS 2010]**
   (a) Matching (b) Blinding
   (c) Randomisation (d) Multivariate analysis

Ans: (b)

8. **Gold standard study for Clinical research is:** **[AIIMS 2011]**
(a) Randomised double-blind trial (b) Systematic meta-analysis
(c) Ecological study (d) Retrospective cohort study

Ans: (b)

9. **About RCT all are true *except*:** **[AIIMS 2011]**
(a) Baseline characteristics are comparable (b) Bias eliminated by double blinding
(c) Sample size depends on type of study (d) Dropouts are excluded from the study

Ans: (d)

10. **Maximum tolerated dose of a new drug is evaluated in:** **[AIIMS 2013]**
(a) Phase 1 (b) Phase 2
(c) Phase 3 (d) Phase 4

Ans: (a)

11. **Selection bias can be eliminated by:** **[DNB December 2011]**
(a) Randomization (b) Single blinding
(c) Double blinding (d) Matching

Ans: (a)

12. **What is not true about cross-sectional study:** **[NEET Pattern 2016]**
(a) Also called prevalence study (b) Snapshot of a population
(c) Shows pattern of disease (d) Tells etiology

Ans: (d)

13. **Incidence rate is calculated by:** **[NEET Pattern 2015]**
(a) Retrospective study (b) Cross sectional study
(c) Prospective study (d) All of the above

Ans: (c)

14. **Odds' ratio is an estimate of:** **[NEET Pattern 2015]**
(a) Relative risk (b) Attributable risk
(c) Prevalence's (d) Incidence rates

Ans: (a)

15. **Regarding case control study true is:** **[NEET Pattern 2014, JIPMER 2014]**
(a) Useful for rare diseases (b) Incidence can be calculated
(c) Takes longer time (d) Relative risk can be calculated

Ans: (a)

16. **Case control studies do not provide following *except*:** **[NEET Pattern 2014]**
(a) Attributable risk (b) Prevalence of disease
(c) Incidence of disease (d) Odds ratio

Ans: (d)

17. Calculate the relative risk for a population in which incidence of disease among exposed is 2(and non exposed is 4: **[NEET Pattern 2012**

(a) 16
(b) 0.5
(c) 24
(d) 5

Ans: (d)

18. Attribute risk gives a better idea of: **[NEET Pattern 2017]**

(a) Strength of association between cause and effect
(b) Impact of successful preventive health programme
(c) Assessing aetiological role or factor in disease
(d) Potential public health importance of disease

Ans: (a)

19. Healthy worker effect is bias of which type? **[NEET Pattern 2013]**

(a) Selection bias
(b) Recall bias
(c) Confounding bias
(d) Berksonian bias

Ans: (a)

20. True about cross-sectional epidemiological study is: **[NEET Pattern 2015]**

(a) Suitable for study of rare diseases
(b) Chronic diseases can be studied
(c) Involves few number of subjects
(d) Relatively inexpensive study

Ans: (b)

21. Bias due to wrong interpretation of laboratory test results and inter-observer variation: **[DNB Pattern 2015]**

(a) Selection bias
(b) Sampling bias
(c) Observation bias
(d) Recall bias

Ans: (c)

Attributes	Ecological	Cross sectional	Case Control	Cohort
Probability of selection bias	NA	Medium	High	Low
Probability of recall bias	NA	High	High	Low
Loss to follow up	NA	NA	Low	High
Confounding	High	Medium	Medium	Medium
Time needed	Low	Medium	Medium	High
Cost	Low	Medium	Medium	High
Strength of causal inference	Low	Low	Medium	High
Investigation of rare disease	++++	-	+++++	-

Contd.

Contd.

Investigation of rare exposure/cause	++	-	-	+++++
Testing multiple outcome of an exposure	+	++	-	+++++
Study of multiple exposure and determinants	++	++	++++	+++
Temporal association	++	-	+	+++++
Incidence rates	-	-	-	+++++
Investigation of long latent periods	-	-	+++	-

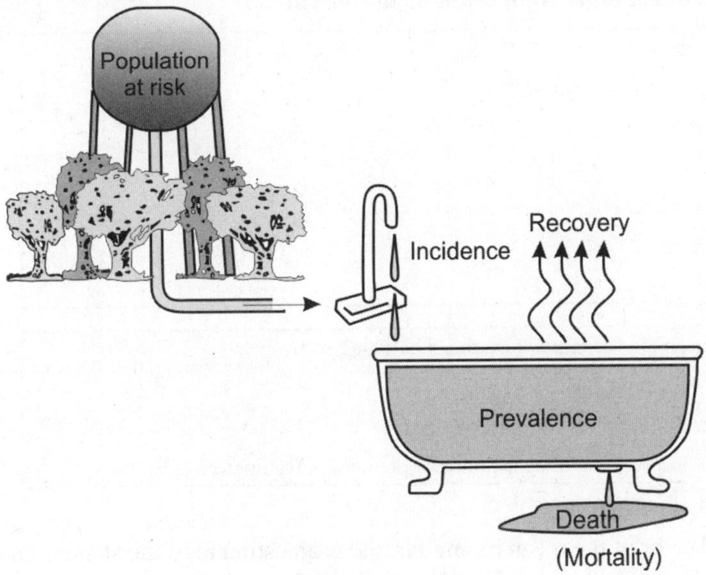

Figure: The epidemiologist's bathtub

PART 4: MULTIPLE CHOICE QUESTIONS

PRACTICE PAPER 1

1. For each of the following questions, select the correct statistical description. Each response may be used once, more than once, or not at all.

(a)	Standard deviation
(b)	Standard error of the mean
(c)	Type I error
(d)	Type II error
(e)	95% Confidence Interval
(f)	Reference range
(g)	Standard error

1.	Sample mean plus or minus 1.96 times its standard error
2.	Standard deviation/square root of n
3.	An estimate of variability of observations
4.	A measure of precision of an estimate of population parameter

Ans: 1.(c) 2.(b) 3.(a) 4.(g)

- Standard error of the mean of one sample is an estimate of the standard deviation that would be obtained from the means of a large number of samples drawn from that population.
- A standard deviation is a sample estimate of population parameter: that is. It is an estimate of the variability of the observations.
- A standard error on the other hand is a measure of precision of an estimate of a population parameter.

2. Match the following:

(a) Chi-square test

(b) Student t-test

(c) Mann-Whitney test

(d) McNemar's test

(e) Spearman rank correlation

(f) Logistic regression

(g) Kruskal-Wallis test

What is the most appropriate statistical test in the following circumstances?

1. Each response may be used once, more than once, or not at all
2. Input variable and outcome variable are bath ordinate
3. Input variable is nominal and outcome variable is non-normal variable is nominal and outcome variable is normal
4. Input variable is nominal and outcome variable is quantitative discrete

Ans: 1.(e) 2.(c) 3.(b) 4.(c)

3. Concerning the p-value, which of the following statements is TRUE?

(a) p-value of 0.001 with significance level set at 0.05 indicates null hypothesis is wrong

(b) For a p-value to be significant it must be a value less than 0.05

(c) A type I error is to reject the null hypothesis when it is true

(d) p-value of 0.0001 indicates a highly significant clinical finding

Ans: (c)

Explanation:

- A type 1 error is to reject the null hypothesis when it is true; this often occurs due to a small sample size.
- A p-value less than the significant level set indicates that in test the null hypothesis should be rejected. This is not the same as saying the null hypothesis is wrong.
- p-value can be set at any level between 0 and 1 and is only arbitrarily set as less than 0.05 in most studies.
- Null hypothesis is rejected if the p-value is smaller than or equal to significance level.

4. Concerning statistical testing which of the following statements is FALSE?

(a) Parametric tests make an assumption about the population distribution

(b) A type 2 error is to accept the null hypothesis when it is false

(c) Specificity relates to the number of true positives in a test

(d) A type 1 error is to reject the null hypothesis when it is true

Ans: (c)

Explanation:

- Sensitivity refers to true positive test results and specificity to true negative results.
- Parametric tests make an assumption about the population distribution and assume a Gaussian or normal distribution.
- Standard error of population is calculated as the standard deviation divided by the square root of the population size.
- A type 1 error is to reject the null hypothesis when it is true and a type 2 error is not rejecting the null hypothesis when it is false.

5. In surgical practice, type 1 error is: **[NEET SS Pattern]**

(a) True positive (b) True negative

(c) False positive (d) False negative

Ans: (c)

6. P value in statistics meaning: **[NEET SS Pattern]**

(a) Probability of rejecting null hypothesis when it is true

(b) Probability of rejecting null hypothesis when it is false

(c) Probability of accepting null hypothesis when it is false

(d) Probability of accepting null hypothesis when it is true

Ans: (a)

7. Median is not used in: **[NEET SS Pattern]**

(a) BP measurements (b) Survival time

(c) Blood group (d) Out of pocket expenditures

Ans: (c)

8. For a screening test, if sensitivity = 90%, specificity = 80% and prevalence = 30%, then PPV of the test is:

(a) 65% (b) 50%

(c) 90% (d) Less than 30%

Ans: (a)

9. In designing a trial comparing primary surgery against primary chemo-radiotherapy for cancer, which of the following statements concerning randomization is TRUE?

(a) The number of patients in each arm will be the same

(b) The clinician can favor one treatment over another if the patient expresses choice

(c) The clinician will not be aware of which treatment the patient has received

(d) The patients in each arm should have similar prognostic factors

Ans: (d)

- Main purpose of randomization is each group will have similar prognostic factors.
- Randomization would need to be restricted to ensure equal numbers in each group.
- The clinician cannot exert bias over the selection group.
- Comparing surgery with chemo-radiotherapy, it will be obvious following treatment which arm the patient was randomized too unless the data is analysed by a blinded researcher.
- Randomization does not ensure that there is no bias between groups.

10. A new screening test for a squamous cell carcinoma is developed. 150 people were tested in total. Within the whole study 50 patients were known to have the disease. When testing was

Started 40 patients that tested positive had the disease; while 10 patients that tested positive did not have the disease. Which of the following statements is CORRECT?

(a) The specificity of the test is 80%

(b) The sensitivity of the test is 80%

(c) The positive predictive value of the test is 90%

(d) The negative predictive value of the test is 80%

Ans: (b)

Following table needs to be constructed to answer this question.

SQUAMOUS CELL CARCINOMA

		Positive	Negative	TOTAL
	Positive	**40**	**10**	**50**
Screening Test	**Negative**	**10**	**90**	**100**
	TOTAL	**50**	**100**	**150**

- The specificity of the test is calculated as the number of true negative results divided by the total number of patients without the disease; 90/100 (90%).
- The sensitivity of the test is calculated as the number of results divided by the total number of patients with the disease; 40/50 (80%).
- The positive predictive value of the test is calculated as the number of true positive results dividedly the total number of positive tests; 40/50 (80%).
- Negative predictive value of the test is calculated as the number of true negative results divided by total number of negative tests; 90/100 (90%).
- Prevalence is calculated as number with disease over sample size: 50/150 (33%).

11. Cross-sectional study:

(a) Subjects are divided into groups on random allocation

(b) Assessment made for a small subset of population at a given time

(c) A large subset of population for long period

(d) Assessment made when one part of population is taken as control

Ans: (b)

- Simplest form of an observational study (Descriptive/Analytical).
- Based on *single examination* of cross-section of sample population at 1 point of time.
- Provides very little information about the natural history of disease or about the rate of occurrence of new cases (incidence).
- Measures *disease and exposure simultaneously* in a well-defined population.

12. Which type of Study is best described in the each response may be used more than once or not at all?

1. A prospective comparison of 2 treatments
2. Retrospective study of patients with disease in an attempt to identify risk for disease
3. A survey to examine the prevalence of a disease
4. A prospective study in which subjects initially disease-free, are followed up

(a) Case control study
(b) Cohort study
(c) Randomized controlled trial
(d) Case series study

Ans: 1.(c) 2.(a) 3.(d) 4.(b)

13. A researcher intending to study possible factors causing lung cancer, visited chest hospital and collected data from patients and their records regarding lung cancer and factors like smoking, blood group, occupation and dietary pattern. Which one of the following best describes study method?

(a) Cohort study
(b) Ecologic study
(c) Cross-sectional study
(d) Case control study

Ans: (c) Cross-sectional study

CROSS-SECTIONAL STUDY:

- This study is based on single examination of a cross-section of a population "at a point of time" results of sample are projected to whole population.
- Provides prevalence of disease under study.
- Can't establish causality.
- Provides little information about natural history of the disease.

PRACTICE PAPER 2

1. **What is arithmetic average of observations belonging to a continuous variable?**
   (a) Mean
   (b) Median
   (c) Mode
   (d) None of the above

Ans: (a)

2. **Which of the following are measures of dispersion around the mean?**
   (a) Standard deviation (SD)
   (b) Variance
   (c) Range
   (d) All of the above

Ans: (d)

3. **What proportion of values lie between 1 SD around mean in normal distribution?**
   (a) 34%
   (b) 68%
   (c) 95%
   (d) 99%

Ans: (b)

4. **Which of the options in Figure is likely to have the smallest SD?**

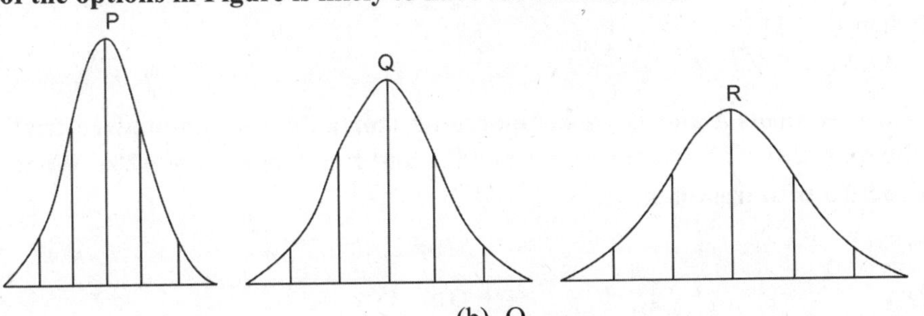

   (a) P
   (b) Q
   (c) R
   (d) All have same SD

Ans: (a)

- P has smallest SD
- Q has greater SD and dispersion

5. **What would be the best test to decide if the mean age of two groups is different?**
   (a) Unpaired t-test
   (b) Paired t-test
   (c) Chi-square test
   (d) Linear regression

Ans: (a)

6. **A group of 40 patients is given a treatment. Measurements of heart rate are taken before and after. What would be the appropriate test to determine if the treatment was effective?**

(a) Unpaired t-test
(b) Paired t-test
(c) Chi-square test
(d) Linear regression

Ans: (b)

7. **Treatment "A" cured 40% of the patients and treatment "B" cured 70% of the patients. What would be appropriate test to determine if 2 treatments are different?**

(a) Unpaired t-test
(b) Paired t-test
(c) Chi-square test
(d) Linear regression

Ans: (c)

8. **Relationship between age and E/A velocity ratio can be studied best with which test?**

(a) Unpaired t-test
(b) Paired t-test
(c) Chi-square test
(d) Linear regression

Ans: (d)

9. **Measures of population description include which of the following?**

(a) Measures of central tendency and dispersion
(b) Mean and median
(c) Inter quartile range and range
(d) Median and SD

Ans: (a)

10. **Effects of treatments A and B are compared on mortality in a randomized trial. At end of 2 years the mortality with treatment A was 15% and treatment B was 12%. What was relative risk reduction with treatment B?**

(a) 3%
(b) 20%
(c) 0%
(d) 25%

Ans: (b)

11. **Which statements regarding subset analysis of trial results is not true?**

(a) All subgroup analysis should be prespecified in a statistical analysis plan
(b) It is used to look at trends in prespecified subsets and interactions
(c) Any deviation from expected trend is hypothesis generating rather than conclusive
(d) When the trial is negative and one of the subsets shows a positive result, it confirms benefits in that subset because it is a randomized trial

Ans: (d)

12. Two treatments for BP reduction are tested. The range of systolic BPs qualifying for trial entry is 160-220 mm Hg. What will be the best way to analyze the results?

(a) Analysis of covariance (ANCOVA) (b) Paired t-test

(c) Unpaired t-test (d) Kaplan-Meir analysis

Ans: (a)

13. Noninvasive diagnostic test to sort out symptom is most productive in which patient groups?

(a) Low probability of disease (b) Intermediate probability of disease

(c) High probability of disease (d) All of these groups

Ans: (b)

14. A group of 100 patients with documented disease and 100 without disease got a diagnostic test to detect that disease. Among those with disease 80 had a positive test, and among those without disease 20 had a positive test. What is sensitivity of the test?

(a) 80% (b) 20%

(c) 40% (d) 50%

Ans: (a)

15. Which of the following test characteristics are ideal for screening of disease?

(a) High sensitivity, low cost, minimal risk (b) High specificity, low cost, minimal risk

(c) Either A or B (d) Neither A nor B

Ans: (a)

16. Which test characteristics are ideal for confirming the presence of a disease?

(a) High sensitivity, low cost, minimal risk (b) High specificity, low cost, minimal risk

(c) Either A or B (d) Neither A nor B

Ans: (b)

17. Which of the following will false positives increase with?

(a) Higher pretest probability (b) Lower pretest probability

(c) Intermediate pretest probability (d) None of the above

Ans: (b)

18. As the test positivity criterion gets more stringent (e.g., increasing ST depression requirement to 2 mm instead of 1 mm), which of the following will happen to the test?

(a) Increase in sensitivity and specificity

(b) Decrease in sensitivity and specificity

(c) Decrease in sensitivity and increase in specificity

(d) Increase in sensitivity and decrease in specificity

Ans: (c)

19. What is positive likelihood ratio (LR+) of test with 90% sensitivity and 90% specificity?

(a) 1 (b) 5

(c) 9 (d) 90

Ans: (c)

20. You want to test if treatment "X" prolongs survival in a disease under study. What would be the best study design?

(a) A double-blind randomized controlled clinical trial with power > 0.8, an outcome adjudication committee, and safety monitoring committee harmed

(b) Single blind trial so that you can monitor to make sure that the patient is not harmed

(c) A smaller trial with outcome surrogates

(d) A non-randomized trial with propensity score matching

Ans: (a)

21. A large registry of patients with aortic stenosis was analyzed to see if aortic valve replacement (AVR) would prolong survival in patients with severe aortic stenosis with low gradient and normal ejection fraction. What would be the best statistical approach?

(a) Kaplan-Meier analysis (b) Cox regression analysis

(c) Propensity score analysis (d) None of the above

Ans: (c)

22. Some of the measures of imprecision or uncertainty around point estimates include which of the following?

(a) 95% confidence interval (b) SD

(c) Standard error of the mean (d) All of the above

Ans: (d)

23. A study is undertaken to determine if prenatal exposure to marijuana is associated with low birth weight in infants. Mothers of 50 infants weighing less than 5 lbs (LBW) and 50 infants weighing more than 7 lbs (normal birth weight) are questioned about their use of marijuana during pregnancy. Study finds that 20 mothers of LBW infants and 2 mothers of normal birth-weight infants used drug during pregnancy. Odds ratio associated with smoking marijuana during pregnancy is:

(a) 10 (b) 16

(c) 20 (d) 30

Ans: (b)

24. A case-control study is done to determine if elderly demented patients are more likely to be injured at home than elderly patients who are not demented. The results of the study show an odds ratio of 3. This figure means that if an elderly patient was injured at home, that patient:

(a) Should be placed in an extended-care facility

(b) Was one-third more likely to be demented than a patient who was not injured at home

(c) Was no more likely to be demented than a patient who was not injured at home

(d) Was three times more likely to be demented than a patient who was not injured at home

Ans: (d)

25. A patient is given a new screening test for TB. Although the patient is infected, the test indicates that the patient is well. This test result is known as:

(a) False positive
(b) False negative
(c) True positive
(d) True negative

Ans: (b)

26. There are 100,000 people in India. On January 1, 2013, 50 of these people have Disease Y. Fifty divided by 100,000 on that date gives which of following measures for Disease Y?

(a) Point prevalence
(b) Period prevalence
(c) Incidence rate
(d) Odds ratio

Ans: (a)

27. A study is done to determine the effectiveness of a new antihistamine. To do this, 25 allergic patients are assigned to one of 2 groups, new drug (13 patients) or placebo (12 patients). Patients are then followed over a 6-month period. Study is best described as:

(a) Cohort study
(b) Cross-sectional study
(c) Case control study
(d) Clinical treatment trial

Ans: (d)

28. A study is designed to determine the relationship between emotional stress and peptic ulcer. To do this, the researchers use hospital records of patients diagnosed with peptic ulcer disease and patients diagnosed with other disorders over period from July 2002-2012. The emotional stress each patient was exposed to was determined from these records and was then quantified with a score of 1 (least stress) to 100 (most stress). This study is best described as a:

(a) Cohort study
(b) Cross-sectional study
(c) Case-control study
(d) Historical cohort study

Ans: (c)

29. Using the data in the table below, chance of surviving for 2 years after being diagnosed with ovarian cancer is 360/500 =72% and chance of surviving for 4 years is 135/500=27%. What is the chance of surviving for 4 years after the original diagnosis, given that the patient is alive at the end of the second year.

Year	Number of women in beginning of year	Number of women who died during year	Percent of women who survived each year (%)
0-1	500	100	80
1-2	400	40	90
2-3	360	90	75
3-4	270	135	50

(a) 40/400 = 10% (b) 135/500 = 27%

(c) 40/135 = 29.6% (d) 135/360 = 37.5%

Ans: (d)

30. A type of gynecological cancer has the same incidence rate in white women and African American women in United States, but prevalence rate of this type of cancer is lower in African American than in white women. Most likely explanation for difference in prevalence rates is that when compared to white women. African-American women are more likely to:

(a) Recover from this type of cancer

(b) Have natural immunity to this type of cancer

(c) Have increased access to treatment for this type of cancer

(d) Die from this type of cancer

Ans: (d)

31. A new blood test to detect prostate cancer by measuring PSA was given to 1,000 male members. Although 50 of men actually had prostate cancer, test was positive (PSA>4 ng/mL) in only 15; other 35 patients with prostate cancer had negative tests. Of the 950 men without prostate cancer, the test was positive in 200 men and negative in 750. Specificity of this test is:

(a) 15% (b) 30%

(c) 48% (d) 79%

Ans: (d)

32. If cut off value indicating a positive test is lowered from PSA 4 ng/mL to PSA 3 ng/mL, this change would:

(a) Increase negative predictive value (b) Decrease sensitivity

(c) Increase false negative rate (d) Increase positive predictive value

Ans: (a)

33. A research study is done to determine if intravenous ibandronate sodium will decrease incidence rate of hip fractures in perimenopausal women. There are 2,600 women in ibandronate sodium group, of whom 130 develop hip fractures. Of 2,600 women in the placebo group, 260 develop hip fractures. Based on these data, how many women need to be treated with ibandronate sodium to prevent one hip fracture?

(a) 1 (b) 5

(c) 10 (d) 20

Ans: (d)

34. A study is designed to compare new medication for Crohn disease with standard medication. To do this, each of 50 Crohn disease patients is allowed to decide which of these 2 treatment groups to join. Major reason that results of this study may not be valid is because of:

(a) Selection bias (b) Recall bias

(c) Sampling bias (d) Differences in the sizes of the two groups

Ans: (a)

35. A blood test reveals that a 35-year-old woman at 18 weeks gestation has increased serum AFP. Of the following measures, which has greatest influence in determining predictive value of this test for neural tube defects in the fetus?

(a) Absolute concentration of AFP in the maternal serum

(b) Family history of dizygotic twin pregnancy

(c) Prevalence of neural tube defects in population

(d) Specificity of the blood test

Ans: (c)

36. After a new antidepressant has been on the market for 5 years, it is determined that of 2,400 people who have taken drug, 360 complained of persistent nausea. If physician has 2 patients on this antidepressant, probability that both of them will experience persistent nausea is approximately:

(a) 2% (b) 9%

(c) 24% (d) 30%

Ans: (a)

37. In people with no known risks for tuberculosis, a positive reaction to PPD tuberculin skin test requires 15 mm or more of hard swelling at the site. A group of physicians decide that they are going to change the criterion for a positive test in a group of people with no known risk for tuberculosis to a hard swelling of 10 mm or more at the site. With respect to the PPD test, this change in the cutoff point is most likely to:

(a) Increase sensitivity (b) Decrease sensitivity

(c) Decrease negative predictive value (d) Increase positive predictive value

Ans: (a)

38. Which of the following is the correct definition of sensitivity?

(a) The strength of the linear association between two variables

(b) The ratio of the odds of disease between two groups

(c) The probability of testing negative when the condition is not present

(d) The probability of testing positive when the condition is present

Ans: (d)

39. Which of the following correctly pairs the probability used with hypothesis testing with its definition?

(a) Alpha (α): 1-probability of a type II error (b) Beta (β): probability of a type I error

(c) Power: probability of a type II error (d) P value: observed significance level

Ans: (d)

40. Which of the following measures of variability is defined as the difference between the 25th and 75th percentiles?

(a) Standard deviation
(b) Standard error
(c) Interquartile range
(d) Confidence interval

Ans: (c)

41. Which of the following is defined as proportion of cases at a given point in time?

(a) Percent
(b) Rate
(c) Prevalence
(d) Incidence

Ans: (c)

42. Which of the following correctly pairs a summary statistic for central tendency with its definition?

(a) Mean: most frequently occurring value
(b) Median: middle value when the data are ordered from smallest to largest
(c) Mode: arithmetic average of a set of numbers
(d) Confidence interval: area of variation between upper and lower limits on particular scale

Ans: (b)

43. Which of the following correctly describes continuous data?

(a) Data that can be any numeric value within a specified range
(b) Data that have a finite number of categories
(c) Data that have a finite number of categories with an implied order
(d) None of the above

Ans: (a)

44. Which of the following correctly describes ordinal data?

(a) Data that can be any numeric value within a specified range
(b) Data that have a finite number of categories
(c) Data that have a finite number of categories with an implied order
(d) Data that can have only one of two specified values

Ans: (c)

45. Which of the following correctly describes categorical data?

(a) Data that can be any numeric value within a specified range
(b) Data that have a finite number of categories
(c) Data that have a finite number of categories with an implied order
(d) Data that can have only one of two specified values

Ans: (b)

46. **A study is conducted to investigate the effectiveness of a new analgesic in relieving headache. A subject's headache intensity is recorded at baseline and subsequently at 15-minute intervals after treatment administration for upto 1 hour. Headache intensity is measured using a 0 to 100 mm visual analog scale with 0 representing no HA and 100 representing worst pain imaginable. Initial analysis consists of comparing various outcomes between treatment group and control group. Which statistical test would you use to measure headache intensity at 30 minutes?**

(a) Pearson's chi-square (χ^2) test (b) Student's t test
(c) Log-rank test (d) Mann-Whitney U test

Ans: (b)

47. **Which study design is most appropriate for conducting a survey on people's use and attitudes toward seat belts?**

(a) Randomized clinical trial (b) Meta-analysis
(c) Case control (d) Observational
(e) Crossover

Ans: (d)

48. **Study design is most appropriate for comparing effectiveness of 2 analgesics in relieving headache pain in patients presenting to emergency department with severe headaches?**

(a) Randomized clinical trial (b) Meta-analysis
(c) Case control (d) Observational
(e) Crossover

Ans: (a)

49. **Which study design is most appropriate for determining factors associated with skeletal fluorosis in sibling pairs (one sibling with skeletal fluorosis and one without)?**

(a) Randomized clinical trial (b) Meta-analysis
(c) Case-control (d) Observational

Ans: (c)

50. **When compared to the 99% confidence interval, the 95% confidence interval is:**

(a) Less precise and less accurate (b) More precise but less accurate
(c) More precise and more accurate (d) Less precise and more accurate

Ans: (b)

PRACTICE PAPER 3

1. **Which of the following statistical tests is most appropriately used to evaluate the difference in percentage of women who lose weight on a protein-sparing diet versus percentage whose weight on high-protein diet?**

   (a) Paired t-test (b) Analysis of variance

   (c) Chi-square test (d) Correlation

Ans: (c)

2. **Which of the following statistical tests is most appropriately used to evaluate differences between initial body weight and final body weight for each woman on protein sparing diet?**

   (a) Paired t-test (b) Analysis of variance

   (c) Chi-square test (d) Correlation

Ans: (a)

3. **Which of the following statistical tests is most appropriately used to evaluate relationship between body weight and systolic blood pressure in a group of 25-year-old women?**

   (a) Paired t-test (b) Analysis of variance

   (c) Chi-square test (d) Correlation

Ans: (d)

4. **In a study to determine the usefulness of a new antihypertensive medication, 12 hypertensive patients are given the new drug and 10 hypertensive patients are given a placebo. Dependent variable in this study is:**

   (a) Experimenter's bias

   (b) Giving the patients the drug

   (c) Giving the patients a placebo

   (d) Patients' blood pressure following treatment with the drug or placebo

Ans: (d)

5. **For a group of 20 elderly patients aged 85 to 90, blood pressure measurements are taken and recorded three times per day over a 2-week period. At the end of the 2 weeks, the patients' blood pressures are taken and recorded four times per day for the next 2 weeks. If the patients' blood pressures fall within a normal distribution, what will be the effect of taking the blood**

pressures four times per day rather than three times per day on standard deviation and shape of the resulting curve?

(a) Standard deviation decreases; shape of the curve does not change

(b) Standard deviation increases; shape of the curve does not change

(c) Standard deviation decreases; shape of the curve is positively skewed

(d) Standard deviation increases; shape of the curve is negatively skewed

Ans: (a)

6. A pharmaceutical company claims that their new drug for genital herpes decreases the persistence of an outbreak from 5.5 days to 5.3 days. This difference is significant at 95% confidence interval. This result:

(a) Is clinically significant but not statistically significant

(b) Is not clinically significant but is statistically significant

(c) Is neither clinically nor statistically significant

(d) Is both clinically and statistically significant

Ans: (b)

7. On a gross anatomy quiz, test scores of 10, 10,10, 70, 40, 20 and 90 are obtained by seven students in a laboratory group. If the class teaching assistant erred and recorded the grade of one student who got 10 as 100, mean, median, and mode would, respectively:

(a) Increase, increase, increase

(b) Increase, not change, not change

(c) Increase, increase, not change

(d) Increase, not change, not change

Ans: (c)

8. On a gross anatomy quiz, test scores of 10, 10,10, 70, 40, 20 and 90 are obtained by seven students in a laboratory group. Which of the following correctly describes quiz scores?

(a) Positively skewed

(b) A normal distribution

(c) Negatively skewed

(d) The mode is higher than the mean

(e) The mode is equal to the mean

Ans: (a)

9. Analysis of the data from a large research study reveals a P value of 0.001. These results indicate that the researcher:

(a) Has committed a type I error

(b) Has committed a type II error

(c) Can reject the null hypothesis

(d) Cannot reject the null hypothesis

Ans: (c)

10. A research study is designed to identify the mean body weight of women between the ages of 30 and 39 in Los Angeles. To do this, a researcher obtains the body weights of an unbiased sample of 81 women in Los Angeles in this age group. Mean body weight of the women in the

sample is 135 pounds with a standard deviation of 18. What is the estimated standard error of the mean for this population?

(a) 0.05 (b) 0.10

(c) 1.0 (d) 2.0

Ans: (d)

11. **Systolic blood pressure is normally distributed with a mean of 120 mm Hg and a standard deviation of 10. What percentage of people in a population selected at random would be expected to have systolic blood pressure at or above 140 mm Hg?**

(a) 1.9% (b) 2.5%

(c) 13.5% (d) 34.0%

Ans: (b)

PART 5: "THINK TANK"

LAST MINUTE REVIEW POINTS

1. Symmetric bell shaped curve of a population distribution is Normal distribution.
2. If a value is chosen from a community then the probability that it will be above the median is 0.5.
3. Best method for collecting vital statistics in India is Active surveillance.
4. The sample registration system (SRS) was stated to get information on birth and death rates for the states and country.
5. 'Lay reporting' involves collection of information lay non-professional health workers.
6. Chi-square * test offers an alternate method of testing the significance of difference between two proportions.
7. Systematic random sample is done by picking every 5th or 10th unit at regular intervals.
8. The statistical analysis of two quantitative data (N=200) is done by "Z test".
9. Collection of data about the health and sickness of a population is primary data.
10. Continuous account of the frequency of disease in community is provided by disease registers.
11. Most satisfactory method of collecting data is oral questionnaires.
12. Sex composition can be demonstrated in simple bar chart.
13. Quantitative data can be represented by Bar diagram.
14. A pictorial diagram of frequency distribution is denoted by histogram.
15. "Line diagrams" are used to show trend of event with passage of time.
16. The percentage is depicted in Pie chart.
17. Best way to study relationship between two variables is "scatter diagram".
18. Central value of a series is termed as average.
19. Median is almost equivalent to 50th percentile.
20. If the data is arranged in descending order, the middle value is median.
21. The most frequently occurring value in a data is mode.
22. In a statistical analysis, dispersion of data is measured by range.
23. The maximally used as measure of dispersion is standard deviation.
24. Root mean square deviation denotes standard deviation.
25. Normal distribution curve depends on mean and standard deviation.
26. In a standard normal curve the area between one standard deviation on either side will be 68% while 2 standard deviation course 95%.
27. Normal curve is symmetrical.

28. Limits set up on either side of mean in a normal curve is called confidence limit.
29. Standard error is due to sampling error.
30. Standard error of means is calculated from standard deviation of means.
31. Significant 'P' value is below 0.05.
32. Square root of pq/n indicates standard error of proportions.
33. Summarising index for discrete variates is rate.
34. In order to find out whether there is significant association or not between two quantitative variables, we usually calculate coefficient of correlation.
35. An average of the squares of the amount by which the individual members deviate from the mean of the groups is called the variance.
36. The statistics used to measure the linear association between two characteristics in the same individual is coefficient of correlation.
37. The term record linkage is used to describe the process of bringing together records relating to one individual (or to one family), the records originating in different times or places.
38. Tabulation is the 1st step before the data is used for analysis or interpretation.
39. A randomized trial comparing efficacy of two regimens showed that difference is statistically significant with P<0.001 but in reality the two drugs do not differ in their efficacy. This is an example of Type I error or a [?] error.
40. A frequency distribution may also be represented diagrammatically by the frequency polygon.
41. Diagrams are better retained in the memory than statistical tables.
42. Pictograms are a popular method of presenting data to the "man in the street" or those who cannot understand orthodox data.
43. In statistical maps, the shaded maps are used to present data of varying size.
44. In cluster sampling, natural group is taken as sampling unit.
45. The cause of death in a village is assessed by samples registration system.
46. While applying chi-square test to a contingency table of 4 rows and 4 columns, the degrees of freedom would be 9.

    NOTE: Df = (r – 1) (c – 1) = (4 – 1) (4 – 1) = 3 × 3 = 9.
47. You have diagnosed a patient clinically as having SLE and ordered six tests. Out of which four tests have come +ve and two are –ve. To determine the probability of SLE at this point, you need to know prior probability of SLE, sensitivity and specificity of each tests.
48. A diagnostic test for a particular disease has a sensitivity of 0.90 and a specificity of 0.80. A single test is applied to each subject in the population in which the diseased population is 30%. The probability that a person, negative to this test, has no disease is 90%.
49. Registration of birth and death with a 6 monthly survey is done in sample registration system.
50. Histogram is used to describe quantitative data of a group of patients.
51. If the birth weight of each of the 10 babies born in a hospital in a day is found to be 2.8 kg, then the standard deviation of this sample will be zero.

52. A physician, after examining a group of patients of a certain disease, classifies the condition of each one as normal, mild, moderate or severe. The scale of measurement that is being adopted for classifying the disease condition here is ordinal.

    NOTE: In ordinal, there is definite order is scale presentation like mild, moderate and severe etc.

53. The area under a normal distribution curve for SD of 3 is 97.5%.
54. In a statistical analysis, Range is used to mention the dispersion of data.
55. The sampling method adopted for Universal Immunization Programme coverage evaluation survey of a district is cluster sampling.
56. Quantitative data can be best represented by HISTOGRAM.
57. Mean Incubation Period of Leprosy is calculated from Median.
58. Statistical tests that are non parametric include:
    - The student's 't' test.
    - Wilcoxon rank sum test.
59. Weight in kg is a continuous variable.
60. Examples of Nominal scale include:
    - Age.
    - Sex.
    - Socio-economic status.
61. Limits of confidence of a hypothesis is determined by level of significance.
62. If the systolic BP in a population has a mean of 130 mm Hg and a median of 140 mm Hg. The distribution is said to be negatively skewed.
63. Ten readings of BP taken before and after treatment is studied by paired 't' test.
64. Continuous quantitative variables are expressed by Histogram and Frequency Polygram.
65. Sample registration system is done once in 6 months.
66. If we reject Null Hypothesis, when actually it is true, it is known as Type I error.
67. The deviation from inference of the truth and the process that leads to such deviation is Bias.
68. In the estimation of statistical probability, Z-score is applicable to Normal distribution.
69. Degree of freedom of 2 by 2 contingency table is 1.
70. When the height and weight is perfectly correlated, then the coefficient of correlation is +1.
71. Recognised examples of Random sampling method is simple random, cluster sampling and stratified sampling but not Quota sampling.
72. Confidence limits can be calculated using Mean and standard error.
73. Median weight of 100 children was 12 kg. The standard deviation was 3. The per cent coefficient of variance is 25%.
74. Regression co-efficient is between –1 to +1.
75. If the ELISA test has a sensitivity of 95% and a specificity of 95%, then the +ve predictive value of the test if it is used to screen a community with a prevalence of AIDS of 55 would be 50.

TABLES

Table 1: Calculating Standard Deviation, Standard Error of the Mean, z Score and Confidence Interval

Measure	Comment	Formula
Standard deviation (S)	Average distance of observations from their mean	$S = \sqrt{\frac{\sum(x - \overline{X})^2}{n - 1}}$
Standard error of the mean (SE)	Estimate of the quality of the sample	$SE = \frac{S}{\sqrt{n}}$
z score (z)	Difference between one score in the distribution and the population mean in units of standard deviation	$z = \frac{(X - \overline{X})}{S}$
Confidence interval (Ci)	Specifies high and low limits of the interval in which the true population mean lies	$Ci = \overline{X} \pm z(SE)$

TWO INDEPENDENT VARIABLES:

- Qualitative (normal)–Chi-square test
- Quantitative (continuous)–Student t test

TWO RELATED GROUPS:

- Qualitative (nominal)–Mc Nemars and Chi-square test
- Quantitative (continuous)–Paired t test

MORE THAN 2 GROUPS:

- Qualitative–Chi-square test
- Quantitative–ANOVA - F

Term	Definition	Calculation	Example of complication
Test sensitivity	Proportion of assays with the genetic change that have a positive test result	True positives/(true positives + false negatives)	Impaired amplification of mutant allele in mosaic setting due to selection
Test specificity	Proportion of assays without the genetic change that have a negative test result	True negatives/(true negatives + false positives)	Technical impairment in MLPA leading to false appearance of deletion will lower test specificity
Clinical sensitivity	Proportion of people with a disease who have a positive test result	True positives / (true positives + false negatives)	Multiple genes involved in a disease may lead to some patients not having a particular mutation or set of mutations
Clinical specificity	Proportion of people without a disease who have a negative test result	True negatives / (true negatives + false positives)	Variable penetrance of disease (i.e., carriers of Diamond-Blackfan anemia mutations without symptoms)

Contd.

Contd.

Positive predictive value	The likelihood that a patient has a disease, given a positive test result	True positives / (true positives + false positives)	Variable penetrance or expressivity can complicate this
Negative predictive value	The likelihood that a patient does not have a disease, given a negative test result	True negatives / (true negatives + false negatives)	Multiple genes involved in a disease may complicate this

COHORT STUDIES

HISTORICAL COHORT STUDIES

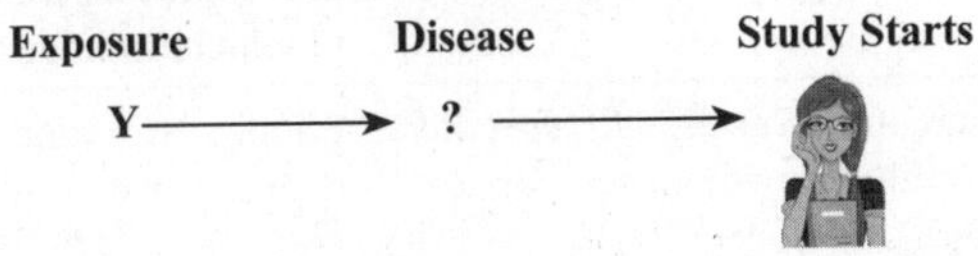

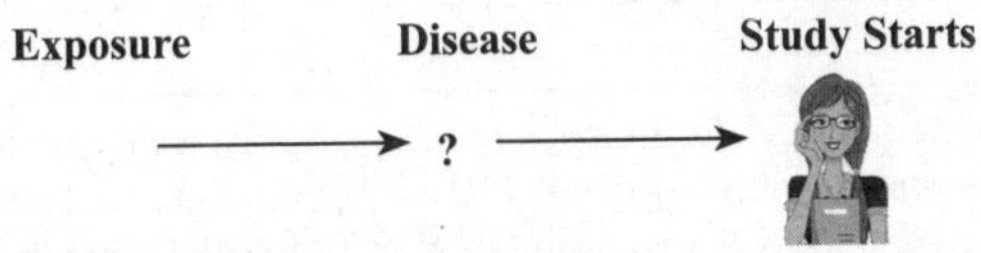

CASE-CONTROL STUDIES

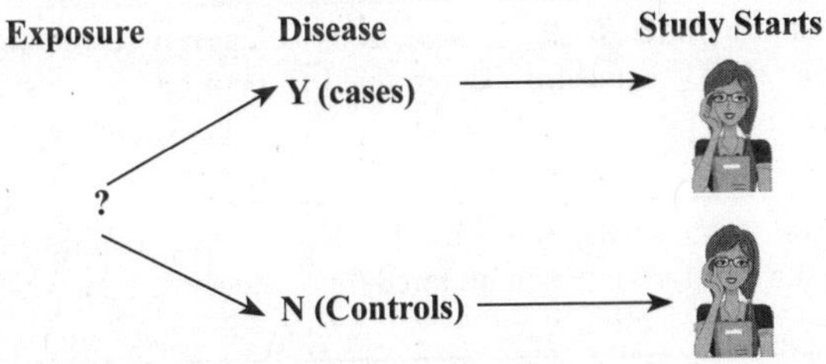

CASE SERIES STUDIES

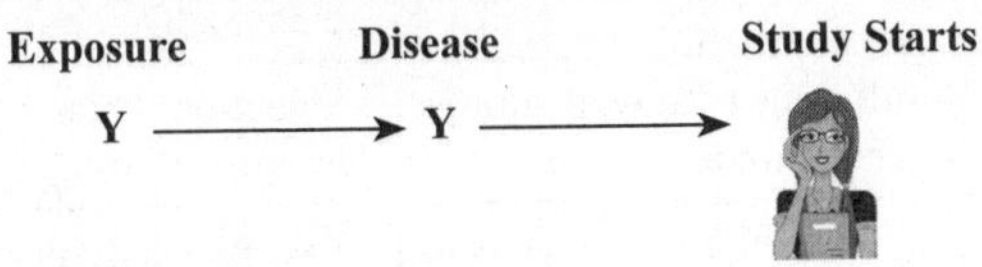

PREVALENCE SURVEYS

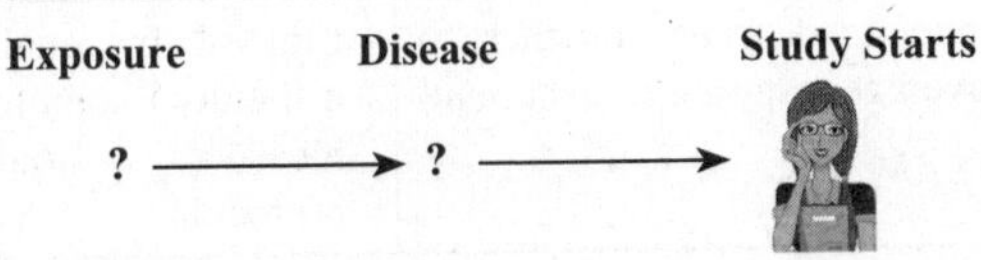

Table 2: Study Designs

	Example	Limitations (Typical)
Descriptive Studies		
Population correlation studies	Rate of disease in population vs. incidence of exposure in population	No link at the individual level, cannot assess or control for other variables. Used for hypothesis generation only
	Changes in disease rates over time	No control for changes in detection technique
Individuals		
Case reports and case series	Identification of rare events, report of outcome of particular therapy	No specific control group or effort to control for selection biases
Cross-sectional surveys	Prevalence of disease in sample, assessment of coincidence of risk factors and disease at a single point in time at an individual level	"Snapshot" view does not allow assessment of causation, cannot assess incident vs. prevalent cases. Sample determines degree to which findings can be generalized
Descriptive cohort studies	Describes outcome over time for specific group of individuals, without comparison of treatments	Cannot determine causation, risk of sample-related biases
Analytic Studies		
Observational case control	Disease state is determined first. Identified control group retrospectively compared with cases for presence of particular risk factor	Highly suspect for bias in selection of control group. Generally can study only one or two risk factors
Retrospective cohort studies	Population of interest determined first, outcome and exposure determined retrospectively	Uneven search for exposure and outcome between groups. Susceptible to missing data, results dependent on entry criteria for cohort
Prospective cohort studies	Exposure status determined in a population of interest, then monitored for outcome	Losses to follow-up over time, expensive, dependent on entry criteria for cohort
Interventional		
Dose escalation studies (phase 1)	Risk for injury from dose escalation	Comparison is between doses, not vs. placebo. Determines toxicity not efficacy.
Controlled nonrandom studies	Allocation to different treatment groups by patient/clinician choice	Selection bias in allocation between treatment groups
Randomized controlled trials	Random allocation of eligible subjects to treatment groups	Expensive. Experimental design can limit generalizability of results
Meta-analysis	Groups randomized trials together to determine average response to treatment	Limited by quality of original studies, difficulty combining different outcome measures. Variability in base study protocols

Table 3: Common Biases in Clinical Research

Bias-Name	Explanation
Sampling Biases	
Prevalence-incidence	Drawing a sample of patients late in a disease process excludes those who have died of the disease early in its course. Prevalent (existing) cases may not reflect the natural history of incident (newly diagnosed) cases
Unmasking	In studies investigating causation, factors that cause symptoms, which is turn cause a more diligent search for the disease of interest. An example might be whether a particular medication caused headaches, which led to the performance of more magnetic resonance imaging studies and resulted in an increase in the diagnosis of arachnoid cysts in patients taking the medication. The conclusion that the medication caused arachnoid cysts would reflect unmasking bias
Diagnostic suspicion	A predisposition to consider an exposure as causative prompts a more thorough search for the presence of the outcome of interest
Referral filter	Patients referred to tertiary care centers are often not reflective of the population as a whole in terms of disease severity and comorbid conditions. Natural history studies are particularly prone to biases of this sort
Chronologic	Patients cared for in previous periods probably underwent different diagnostic studies and treatments. Studies with historical controls are at risk
Nonrespondent/volunteer	Patients who choose to respond or not respond to surveys or follow-up assessments differ in tangible ways. Studies with incomplete follow-up or poor response rates are prone to this bias
Membership bias	Cases or controls drawn from specific self-selected groups often differ from the general population. The result of this bias is the assumption that the group's defining characteristic is the cause of their performance with respect to a risk factor
Intervention Biases	
Co-intervention	Patients in an experimental or control group systematically undergo an additional treatment not intended by the study protocol. If a treatment were significantly more painful than a control procedure, a potential co-intervention would be increased analgesic use postoperatively. A difference in outcome could be due to either the treatment or the co-intervention
Contamination	When patients in the control group receive the experimental treatment, the potential differences between the groups are masked
Therapeutic personality	In unblinded studies, the belief in a particular therapy may influence the way in which the provider and patient interact and alter the outcome
Measurement	
Expectation	Prior expectations about the results of an assessment can substitute for actual measurement in assessments of either diagnosis or therapy, belief in the predictive ability or therapeutic efficacy increases the likelihood that a positive effect will be measured. Unblinded studies and those with a subjective outcome measure are prone to this bias

Contd.

Contd.

Recall	In cohort and case-control studies, different assessment technique or frequencies applied to those with the outcome of interest may increase the likelihood of detection of a risk factor, in particular, asking cases multiple times vs. controls improves chances for recall. This bias applies especially to retrospective studies and is the inverse of the diagnostic-suspicion bias (see earlier)
Unacceptability	Measurements that are uncomfortable, either physically or mentally, may be avoided by study subjects
Unpublished scales	The use of unpublished outcome scales has been shown in certain settings to result in a greater probability of the study finding in favour of experimental treatment
Analysis	
Post hoc significance	When decisions regarding levels of significance or statistical measures to be used are determined after the results are analyzed, it is more likely that a significant result will be found. It is very unlikely that authors would state this in a manuscript, but protection is afforded by publication of the study methods in advance of the study itself
Data dredging	Asking multiple questions of a dataset until something with statistical significance is found. Subgroup analysis, depending on whether preplanned and coherent with the study aims and hypotheses, may also fall into this category

Study#	**Types of bias**	**Explanation**	**Reason for flashless study failure is that when compared to the general population, the subjects**
1	Selection	Rather than making random assignment, the subjects of the investigators are permitted to choose whether an individual will go into the drug group or the placebo group	Who had more severe symptoms to start with chose or were chosen to take flashless rather than placebo and so also had more severe symptoms at the end of the study
2	Sampling	Subjects who volunteer to be in a study are not representative of the population being studied because factors unrelated to the subject of the study (e.g., money) have led them to volunteer	Only joined the study because they needed the money and as such did not represent the typical population of women using drugs such as Flashless
3	Recall	Knowledge of the presence of a disorder alters the way the subject remembers his or her history	Seemed to have more severe menopausal symptoms because they were asked about their symptoms to be in the study
4	Lead-time	Early detection of disease is confused with increased survival or length of time that the symptoms have been present	Seemed to have had menopausal symptoms for a longer period of time because they were identified early to be in the study

Contd.

Contd.

5	Surveillance	People who are aware that they are being followed for the development of a disease are more likely to seek testing for and thus to be identified with the disease	Went to the doctor more often because they were in the study which increased the likelihood of being diagnosis with menopausal symptoms
6	Late-look	People who are most ill are not included in the sample	Who had severe menopausal symptoms chose not to participate in the study, so those who did participate had few symptoms

Matching the strongest research designs to clinical questions	
Question	**Design**
Causation and risk factors	Case-control studies, Cohort studies
Prevalence and diagnosis	Cross-sectional studies
Incidence and prognosis	Cohort studies
Treatment	Randomized controlled trials (RCT)